Regenerative Phenomena

Regenerative
Phenomena

J. F. C. KINGMAN
Professor of Mathematics
in the University of Oxford

John Wiley & Sons Ltd,   London · New York · Sydney · Toronto

Library of Congress Catalog Card Number: 70–39143
ISBN: 0 471 47905 5

Text set in 10/12 pt. Monotype Times Roman, printed
by letterpress, and bound in Great Britain at The
Pitman Press, Bath

for Valerie

Preface

The theory of Markov chains with a continuous time parameter and a countable infinity of states grew, in the hands of Kolmogorov, Lévy, Doob, Feller and others, into a rich mathematical structure. The classical phase of its development may be said to have ended in 1960, when Chung published his magisterial account [6]. Much of the theory there described is concerned with the transition probabilities $p_{ij}(t)$ (where i and j run over the set S of possible states, and t over the positive real numbers), and consists of consequences of the conditions

$$p_{ij}(t) \geqslant 0, \qquad \sum_{j \in S} p_{ij}(t) = 1,$$

$$p_{ij}(s + t) = \sum_{k \in S} p_{ik}(s)p_{kj}(t)$$

which they evidently satisfy, together with some weak assumption of continuity in t. From these postulates alone, though by arguments of strong probabilistic flavour, flow deep results such as Ornstein's theorem that the functions p_{ij} are continuously differentiable.

In any active branch of mathematics, there comes a time when one should take stock, and ask why its theorems are true, and why its unsolved problems are difficult. In this case, the question resolves itself into a more concrete one: what functions of a positive real variable t can arise as transition probabilities $p_{ij}(t)$ in Markov chains? To this inquiry a great part of this book is devoted, and the answer is given in Chapter 6. It is quite hard work to reach the final result, but at various points in the ascent there are revealing views of neighbouring summits which may be as rewarding as the peak itself.

One general strategy has guided the research described here. If, for example, interest centres on the function p_{ii} for a particular state i, the rule is to concentrate on that state, and not to worry too much about the others. Indeed, for many purposes surprisingly little information is lost if the other states $j \neq i$ are lumped together in a composite state 'not i'. The result is a two-state process which is not usually Markovian, but has nevertheless a simple stochastic structure. It is called a regenerative phenomenon.

The very important concept of regeneration goes back to the work of authors such as Palm and Doeblin, and has proved to be of great importance in pure and applied probability. The idea is that, at certain random time instants, a random process may begin again, forgetting its past and moving forward to a new life. Such regeneration points have usually been thought of as rather rare (finitely many in each finite interval) but this is inadequate for the Markov application. The generalisation appropriate to a situation in which regeneration points abound proves to be just the concept of a regenerative phenomenon.

Formally, the book is accessible to a reader familiar with the basic concepts and tools of probability theory as set out for instance in [54] (together with the important theorem of Tychonov [27]). But the motivation may be somewhat obscure to one unfamiliar with the broad outlines of the classical theory of Markov chains [18], [6], and such a reader may also fail to appreciate how far the arguments used here derive historically from those displayed in that theory. An important technical device which appears many times in the proofs is that of weak convergence of measures (on the real line); the necessary theory is described in an appendix which may be looked upon as a case study in the use of this tool.

My own interest in these problems began in the Post Office Engineering Research Station at Dollis Hill, and has continued in the quieter academic groves of Cambridge, Perth, Brighton, Stanford and Oxford. To all those with whom I have discussed the theory, I offer most grateful thanks, and especially to David Kendall, whose insight and scholarship have had a profound influence, and whose friendship has been a continued encouragement.

I have also been fortunate to count myself among the friends of Rollo Davidson, who before his tragic death in the summer of 1970 had made distinguished contributions to this, as to other areas of probability theory. Had he lived, several problems described in these pages as unsolved would not long have remained so.

Oxford, 1971   J.F.C.K.

Note

The final section of each chapter consists of *Notes*, which may be comments on theorems or proofs, examples, counter-examples, unsolved problems, or references to further work in the literature. Where assertions are made with neither justification nor reference to the literature, the reader who so wishes may regard them as exercises, and as such will not find them too difficult.

Theorems are numbered within chapters, so that Theorem A.B is the Bth theorem of Chapter A. An index of theorems will be found on pages 185 and 186. Equations are numbered within sections. Thus in Section A.B, a reference to equation (c) pertains to the same section, a reference to equation (b.c) means equation (c) of Section A.b, while one to equation (a.b.c) refers to equation (c) of Section a.b. References to the literature appear in brackets [ ], and refer to the bibliography on pages 179–184. A supplementary bibliography, added in proof, records some relevant papers which appeared while the book was in the hands of the printer, and which are not therefore mentioned in the text.

The notation is, I hope, in fairly good agreement with that at present customary in probability theory; $\mathbf{P}$ and $\mathbf{E}$ denote probability and expectation respectively. Open and closed intervals are distinguished by parentheses ( ) and brackets [ ] respectively. The Laplace transform of a function f is written $\hat{f}$. The symbol $\blacklozenge$ denotes the end of a proof.

Contents

CHAPTER 1

Regenerative Phenomena in Discrete Time

1.1 MARKOV CHAINS

By a Markov chain in discrete time* we shall mean a sequence $X = (X_0, X_1, X_2, \ldots)$ of random variables, taking values in a countable state space S, and having the property that, for any n, and any $j \in S$,

(1) $$\mathbf{P}(X_n = j \mid X_0, X_1, \ldots, X_{n-1}) = \mathbf{P}(X_n = j \mid X_{n-1}).$$

It will moreover be assumed without further comment that the (one-step) transition probability

(2) $$p_{ij} = \mathbf{P}(X_n = j \mid X_{n-1} = i)$$

does not depend on n.

Under these conditions the joint distributions of the X_n are completely determined by the transition probabilities p_{ij} $(i, j \in S)$ and the initial distribution

(3) $$p_i = \mathbf{P}(X_0 = i);$$

for $i_0, i_1, \ldots, i_n \in S$,

(4) $$\mathbf{P}(X_0 = i_0, X_1 = i_1, \ldots, X_n = i_n) = p_{i_0} \prod_{r=1}^{n} p_{i_{r-1} i_r}.$$

The numbers p_i, p_{ij} satisfy the conditions

(5) $$p_i \geqslant 0, \qquad \sum_{j \in S} p_j = 1,$$

(6) $$p_{ij} \geqslant 0, \qquad \sum_{j \in S} p_{ij} = 1,$$

---

* In matters of notations and terminology we shall usually follow Chung [6]. Thus a Markov chain always has a countable number of possible states, and its time parameter may be discrete or continuous. The conditional probabilities (1) and (2) are to be understood in the elementary sense.

1

and conversely the Daniell–Kolmogorov theorem [54] shows that there is a stochastic process satisfying (1)–(4) whenever (5) and (6) are satisfied.

It is convenient to denote by $\mathbf{P}_k$ $(k \in S)$ the probability measure conditional on $\{X_0 = k\}$, so that

$$(7) \qquad \mathbf{P} = \sum_{k \in S} p_k \mathbf{P}_k.$$

From (4),

$$(8) \qquad \mathbf{P}_k(X_1 = i_1, \ldots, X_n = i_n) = p_{k i_1} \prod_{r=2}^{n} p_{i_{r-1} i_r},$$

so that $\mathbf{P}_k$ depends only on (p_{ij}), and not on the initial distribution (p_i).

If (8) is summed over all values of $i_1, i_2, \ldots, i_{n-1}$, we obtain the n-step transition probabilities

$$(9) \qquad p_{ij}^{(n)} = \mathbf{P}_i(X_n = j),$$

which are determined by (p_{ij}) and are most easily generated using the Chapman–Kolmogorov equation

$$(10) \qquad p_{ij}^{(m+n)} = \sum_{k \in S} p_{ik}^{(m)} p_{kj}^{(n)} \qquad (m, n \geqslant 1, \, i, j, \in S),$$

where of course

$$p_{ij}^{(1)} = p_{ij}.$$

A number of authors, notably Doeblin [14], had used the idea of considering the return of the sequence X_n to a fixed state a before Feller gave his brilliant systematic development of the theory of recurrent events in [17] and [18]. Suppose that $X_0 = a$ (that is, work with the probability measure $\mathbf{P}_a$) and consider, for any realisation of the random sequence X, the set of integers n for which $X_n = a$. There may be a finite or infinite number of them, and they may be written in ascending order as

$$0 = \nu_0 < \nu_1 < \nu_2 < \ldots < \nu_\kappa,$$

where $\kappa \leqslant \infty$ is the total number of returns of the process to its initial state a. For $n \geqslant 1$,

$$\mathbf{P}_a(\kappa \geqslant 1, \nu_1 = n) = \mathbf{P}_a(X_1, X_2, \ldots, X_{n-1} \neq a, X_n = a)$$
$$= \sum_{i_1, i_2, \ldots, i_{r-1} \neq a} p_{a i_1} p_{i_1 i_2} \cdots p_{i_{n-1} a}$$
$$= f_n,$$

say. Clearly $f_n \geqslant 0$ and

$$(11) \qquad \sum_{n=1}^{\infty} f_n = \mathbf{P}_a(\kappa \geqslant 1) \leqslant 1.$$

It is convenient to define $\nu_1 = \infty$ if $\kappa = 0$, so that

$$f_n = \mathbf{P}_a(\nu_1 = n)$$

and

$$(12) \qquad f_\infty = 1 - \sum_{n=1}^{\infty} f_n.$$

The important point is that the 'diagonal' transition probabilities $p_{aa}^{(n)}$ may be computed in terms of the 'recurrence time' probabilities f_n. If we write

$$(13) \qquad u_n = p_{aa}^{(n)},$$

with the convention that $u_0 = 1$, then for $n \geqslant 1$,

$$u_n = \mathbf{P}_a(X_n = a)$$

$$= \sum_{r=1}^{n} \mathbf{P}_a(X_n = a, \nu_1 = r)$$

$$= \sum_{r=1}^{n} f_r \mathbf{P}_a(X_n = a | \nu_1 = r).$$

Since the event $\{\nu_1 = r\}$ depends only on $X_1, X_2, \ldots, X_r$ and implies $\{X_r = a\}$, (1) gives

$$\mathbf{P}_a(X_n = a | \nu_1 = r) = \mathbf{P}(X_n = a | X_r = a)$$

$$= p_{aa}^{(n-r)} = u_{n-r}.$$

Hence

$$(14) \qquad u_0 = 1, \qquad u_n = \sum_{r=1}^{n} f_r u_{n-r} \ (n \geqslant 1).$$

Using (14) as a recurrence relation, the u_n may be written explicitly in terms of the f_n:

$$(15) \qquad
\begin{cases}
u_0 = 1, \\
u_1 = f_1, \\
u_2 = f_2 + f_1^2, \\
u_3 = f_3 + 2f_1 f_2 + f_1^3, \\
u_4 = f_4 + 2f_1 f_3 + 3f_1^2 f_2 + f_2^2 + f_1^4, \\
\cdots
\end{cases}$$

Alternatively, (14) may be summarised in an identity between the generating functions

$$F(z) = \sum_{n=1}^{\infty} f_n z^n, \qquad U(z) = \sum_{n=0}^{\infty} u_n z^n,$$

which takes the form

(16) $$U(z) = \{1 - F(z)\}^{-1},$$

and is valid in $|z| < 1$.

There are problems (cf. §1.6(i)) in which the probabilities f_n are of interest, and can be derived from the u_n by inverting (16) to give

$$F(z) = 1 - \{U(z)\}^{-1},$$

or

(17) $$\begin{cases} f_1 = u_1, \\ f_2 = u_2 - u_1^2, \\ f_3 = u_3 - 2u_1 u_2 + u_1^3, \\ f_4 = u_4 - 2u_1 u_3 + 3u_1^2 u_2 - u_2^2 - u_1^4, \\ \cdots \end{cases}$$

1.2 RENEWAL SEQUENCES

The argument of the previous section shows that the sequence

$$u_n = p_{aa}^{(n)} \qquad (n = 0, 1, 2, \ldots)$$

of diagonal transition probabilities corresponding to the state a has a very special property; it is generated by a recurrence relation (1.14), where the numbers f_n satisfy

(1) $$f_n \geqslant 0, \qquad \sum_{n=1}^{\infty} f_n \leqslant 1.$$

In other words, if we compute the numbers f_n from (1.17), they must satisfy (1), just because the f_n have the probabilistic interpretation

$$f_n = \mathbf{P}(\nu_1 = n).$$

If $f = (f_1, f_2, \ldots)$ is any sequence satisfying (1), the sequence $u = (u_0, u_1, u_2, \ldots)$ defined by the recurrence relation (1.14) is called the *renewal sequence associated with f*. A sequence is a *renewal sequence* if it is the renewal sequence associated with some sequence f satisfying (1). Thus the diagonal transition probabilities corresponding to any state in any Markov chain form a renewal sequence. It is an important remark, attributed [18] to Chung, that any renewal sequence can arise in this way.

Theorem 1.1. *If* $u = (u_n; n \geqslant 0)$ *is any renewal sequence, there exists a Markov chain X and a state a such that, for all n,*

(2) $$u_n = \mathbf{P}(X_n = a | X_0 = a).$$

Proof. Write

$$g_n = 1 - \sum_{m=1}^{n} f_m,$$

and let $N \leqslant \infty$ be defined by

$$N = \sup \{n; g_n > 0\}.$$

Take the state space S to consist of the integers j in $0 \leqslant j \leqslant N$, and the transition probabilities as

$$\begin{aligned} p_{ij} &= g_{i+1}/g_i & \text{if } j = i + 1, \\ &= f_{i+1}/g_i & \text{if } i = 0, \\ &= 0 & \text{otherwise.} \end{aligned}$$

Then, with $a = 0$,

$$\mathbf{P}_0(\nu_1 = n) = \mathbf{P}_0(X_1 = 1, X_2 = 2, \ldots, X_{n-1} = n - 1, X_n = 0)$$

$$= \frac{g_1}{g_0} \frac{g_2}{g_1} \cdots \frac{g_{n-1}}{g_{n-2}} \frac{f_n}{g_{n-1}} = f_n,$$

so that

$$p_{00}^{(n)} = u_n. \qquad \blacklozenge$$

Thus the diagonal case of the theory of Markov transition probabilities is essentially the theory of renewal sequences. A good deal is known about such sequences, and the following theorems are typical.

Theorem 1.2. *If u and v are renewal sequences, and k is a positive integer, then $^k u$ and uv are renewal sequences, where*

(3) $$(^k u)_n = u_{nk},$$

(4) $$(uv)_n = u_n v_n.$$

Proof. According to Theorem 1.1, there exists a Markov chain X satisfying (2). Then $(X_0, X_k, X_{2k}, \ldots)$ is also a Markov chain, in which the renewal sequence corresponding to the state a is $^k u$, since

$$\mathbf{P}_a(X_{nk} = a) = u_{nk} = {}^k u_n.$$

There also exists a Markov chain Y, on a state space S' and independent of X, and a state $b \in S'$, such that

$$\mathbf{P}_b(Y_n = b) = v_n.$$

Then

$$Z_n = (X_n, Y_n)$$

defines a Markov chain on $S \times S'$, in which the renewal sequence corresponding to (a, b) is uv, since

$$\mathbf{P}_{(a,b)}(X_n = a, Y_n = b) = u_n v_n. \qquad \blacklozenge$$

The second assertion of the theorem implies that the set $\mathscr{R}$ of all renewal sequences is a commutative semigroup under the operation of pointwise multiplication. This semigroup has the identity

(5) $$e = (1, 1, 1, \ldots)$$

corresponding to $f_1 = 1, f_n = 0 \ (n \geqslant 2)$. It also has a simple topological structure.

Theorem 1.3. *Let $\mathscr{R}$ be given the topology which it inherits as a subspace of the product space $[0, 1]^\infty$ of all sequences $(x_n; n \geqslant 0)$ with $0 \leqslant x_n \leqslant 1$. Then $\mathscr{R}$ is a compact metrisable space, and u_n and f_n are continuous functions on $\mathscr{R}$.*

Proof. A countable product of metric spaces is metrisable, and by Tychonov's theorem a product of compact spaces is compact. Thus $[0, 1]^\infty$ is compact and metrisable, and its subspace $\mathscr{R}$ is metrisable. Let $j_n \colon [0, 1]^\infty \to [0, 1]$ denote the nth coordinate function (by definition continuous); then the restriction u_n of j_n to $\mathscr{R}$ is continuous.

Define functions $\phi_n (n \geqslant 1)$ on $[0, 1]^\infty$ by the recurrence relation

$$\phi_n = j_n - \sum_{r=1}^{n-1} \phi_r j_{n-r};$$

these are continuous by induction. Comparing this relation with (1.4), we see that, for $u \in \mathscr{R}$,

$$\phi_n(u) = f_n,$$

so that f_n is continuous.

The subset

$$R = \{x; \phi_n(x) \geqslant 0, n = 1, 2, \ldots\}$$

is closed in $[0, 1]^\infty$, and is therefore compact. Clearly $\mathcal{R} \subseteq R$, and the theorem will be proved if we can show that $\mathcal{R} = R$. To do this, take $u \in R$ and write $f_n = \phi_n(u)$. Then $f_n \geqslant 0$ and we need only prove that

$$(6) \qquad \sum_{n=1}^{\infty} f_n \leqslant 1.$$

To do this, we note that, for $0 < z < 1$,

$$U(z) = \sum_{n=0}^{\infty} u_n z^n$$

is finite, and

$$U(z) = 1 + \sum_{n=1}^{\infty} \left(\sum_{r=1}^{n} f_r u_{n-r} \right) z^n$$

$$= 1 + \sum_{m=0}^{\infty} \sum_{r=1}^{\infty} f_r u_m z^{m+r}$$

$$\geqslant U(z) \sum_{r=1}^{\infty} f_r z^r.$$

Hence

$$\sum_{r=1}^{\infty} f_r z^r \leqslant 1,$$

whence (6) follows on letting $z \to 1$ from below. ◆

1.3 ERGODIC THEOREMS

The theory described so far is essentially that of Feller [17], whose book [18] contains a more leisurely account with a rather different emphasis. The most striking achievement of the theory was that it opened the door to a powerful and systematic treatment of the distribution of X_n for large n, part of which will be described in this section.

Theorem 1.4. *Let u be any renewal sequence. Then, for all m, n $\geqslant 0$,*

$$(1) \qquad u_m u_n \leqslant u_{m+n} \leqslant u_m u_n + 1 - u_m.$$

If $u \neq (1, 0, 0, \ldots)$, there exists a unique positive integer d such that

    (i) $u_n = 0$ *unless d divides n,*

    (ii) $u_{rd} > 0$ *for all but a finite number of positive integers r. The limit*

$$(2) \qquad \rho = \lim_{r \to \infty} (u_{rd})^{1/rd}$$

exists in $0 < \rho \leqslant 1$, and the sequence $\bar{u} = (u_n \rho^{-n})$ is a renewal sequence; in particular

(3)
$$u_n \leqslant \rho^n.$$

Proof. Construct a Markov chain with

$$p_{aa}^{(n)} = u_n$$

for some state a; then

$$u_{m+n} - u_m u_n = \sum_{k \in S} p_{ak}^{(m)} p_{ka}^{(n)} - p_{aa}^{(m)} p_{aa}^{(n)}$$

$$= \sum_{k \neq a} p_{ak}^{(m)} p_{ka}^{(n)}.$$

Hence

$$0 \leqslant u_{m+n} - u_m u_n \leqslant \sum_{k \neq a} p_{ak}^{(m)} = 1 - u_m,$$

and (1) is proved.

For any positive integer n, write

(4)
$$\rho_n = \liminf_{r \to \infty} (u_{rn})^{1/rn}.$$

Let m and n be positive integers with highest common factor $k = (m, n)$, and suppose that $u_m > 0$. There exists positive integers a, b with $am - bn = k$. For any integer r, write $h = h(r)$ for the integral part of ar/n. Then

$$h > (ar/n) - 1 = (bn + k)(r/mn) - 1$$

$$= \frac{br}{m} + \left(\frac{kr}{mn} - 1\right) \geqslant \frac{br}{m},$$

so long as $r \geqslant mn/k$, so that the parentheses in the identity

$$rk = (ar - hn)m + (hm - br)n$$

are non-negative integers. Repeated application of the first inequality of (1) gives

$$u_{rk} \geqslant u_m^{ar - hn} u_n^{hm - br}$$

so long as $r \geqslant mn/k$. Now take (rk)th roots and let $r \to \infty$, using the fact that

$$h/r \to a/n$$

and the positivity of u_m, to give

$$\rho_k \geqslant u_n^{1/n}.$$

Hence we have proved that

(5)
$$u_n \leqslant \rho_k^n$$

whenever there exists m with $(m, n) = k$ and $u_m > 0$. In particular, taking $m = n$, we have

$$u_n \leqslant \rho_n^n$$

for all n, the inequality being trivially satisfied if $u_n = 0$. Since $u \neq (1, 0, 0, \ldots)$, there are values of n with $\rho_n > 0$; write

$$d = \min \{n; \rho_n > 0\}, \qquad \rho = \rho_d.$$

Then (ii) follows from (4) with $n = d$.

To prove (i), suppose to the contrary that there exists m with $u_m > 0$ and $(m, d) = \delta < d$. By hypothesis $\rho_\delta = 0$, and (5) with $k = \delta$ shows that $u_{pd} = 0$ for all primes p not dividing m. Since there are infinitely many such p, this contradicts (ii), and the contradiction establishes (i).

For any r, there exists s with $(r, s) = 1$ which is so large that $u_{sd} > 0$. Putting $m = rd$, $n = sd$ in (5), we have

$$u_{rd} \leqslant \rho_d^{rd} = \rho^{rd},$$

and (3) is proved. Together with (4) $(n = d)$, this proves (2).

If u is associated with the sequence f, write

$$\bar{f}_n = f_n \rho^{-n}, \qquad \bar{u}_n = u_n \rho^{-n}.$$

Then $\bar{u}_0 = 1$ and, for $n \geqslant 1$,

$$\bar{u}_n = \rho^{-n} \sum_{r=1}^{n} f_r u_{n-r} = \sum_{r=1}^{n} \bar{f}_r \bar{u}_{n-r},$$

so that $\bar{u} \in [0, 1]^\infty$ and

$$\phi_n(\bar{u}) = \bar{f}_n \geqslant 0.$$

Hence $\bar{u}$ belongs to the set R, which was shown in the proof of Theorem 1.3 to be indentical to $\mathcal{R}$. ◆

The integer d is called the *period* of u; if $d = 1$, u is said to be *aperiodic*. By Theorem 1.2, the sequence

$$^d u = (u_0, u_d, u_{2d}, \ldots)$$

is an aperiodic renewal sequence. For this reason it is usually necessary only to consider aperiodic renewal sequences. Similarly, the final assertion of the theorem means that it is only necessary to consider renewal sequences with $\rho = 1$.

Since $u_n \geqslant f_n$, we have $f_n = 0$ unless d divides n. Conversely, if $f_n = 0$ unless δ divides n, then (1.14) easily implies that $u_n = 0$ unless δ divides n, so that δ divides d. Thus d is the highest common factor of

$$\{n \geqslant 1; f_n > 0\}.$$

It was noted in §1.1 that the variable v_1 might take the value $+\infty$, with probability

$$f_\infty = 1 - \sum_{n=1}^{\infty} f_n.$$

This possibility can be recognised in terms of the associated renewal sequence.

Theorem 1.5. *The probability f_∞ is non-zero if and only if Σu_n converges, and then*

$$(6) \qquad \sum_{n=0}^{\infty} u_n = f_\infty^{-1}.$$

Proof. For $0 < z < 1$,

$$\sum_{n=0}^{\infty} u_n z^n = \left\{ 1 - \sum_{n=1}^{\infty} f_n z^n \right\}^{-1};$$

using the monotone convergence theorem as $z \to 1$,

$$\sum_{n=0}^{\infty} u_n = \left\{ 1 - \sum_{n=1}^{\infty} f_n \right\}^{-1} = f_\infty^{-1}. \qquad \blacklozenge$$

These results are fairly superficial, and the deep theorem in this area is the celebrated Erdös–Feller–Pollard limit theorem [16], which asserts the existence of the limit of u_n as $n \to \infty$ through multiples of d. It has several different proofs, none of which is really simple. The one given here is the discrete form of an argument used in [38] and is closely related to a general technique of Feller and Orey [20]. It first requires a lemma which has some interest of its own.

Lemma. *If u is an aperiodic renewal sequence, there exists a number u_∞ in $0 \leqslant u_\infty \leqslant 1$ and a non-negative integrable function g on $(0, \pi)$, such that*

$$(7) \qquad u_n = u_\infty + \int_0^\pi g(\theta) \cos n\theta \, d\theta.$$

Proof. Regard the identity

$$U(z) = \{1 - F(z)\}^{-1}$$

as holding between two functions

$$U(z) = \sum_{n=0}^{\infty} u_n z^n, \qquad F(z) = \sum_{n=1}^{\infty} f_n z^n$$

of a complex variable z in $|z| < 1$. For $0 < r < 1$,

$$\text{Re}\,\{U(r\,e^{i\theta})\} = 1 + \sum_{n=1}^{\infty} u_n r^n \cos n\theta,$$

and the Fourier inversion formula gives

$$u_n r^n = \frac{2}{\pi} \int_0^{\pi} \text{Re}\,\{U(r\,e^{i\theta})\} \cos n\theta \, d\theta, \qquad (n \geqslant 1)$$

$$1 = \frac{1}{\pi} \int_0^{\pi} \text{Re}\,\{U(r\,e^{i\theta})\} \, d\theta.$$

Thus, for all $n \geqslant 0$,

$$u_n r^n = \int_0^{\pi} g_r(\theta) \cos n\theta \, d\theta,$$

where

$$g_r(\theta) = \frac{2}{\pi} \text{Re}\,\{U(r\,e^{i\theta}) - \tfrac{1}{2}\}.$$

But

$$\text{Re}\,\{U(z) - \tfrac{1}{2}\} = \text{Re}\left\{\frac{1}{1 - F(z)} - \tfrac{1}{2}\right\}$$

$$= \tfrac{1}{2}\,\text{Re}\left\{\frac{1 + F(z)}{1 - F(z)}\right\}$$

$$= \tfrac{1}{2}\,\frac{1 - |F(z)|^2}{|1 - F(z)|^2} \geqslant 0,$$

since $|F(z)| \leqslant F(|z|) \leqslant F(1) \leqslant 1$. Hence g_r is the density of a probability measure γ_r on $[0, \pi]$. The theorems of the Appendix show that, as $r \to 1$, γ_r converges weakly to a probability measure γ, and that

(8) $$u_n = \int_0^{\pi} \cos n\theta \, \gamma(d\theta).$$

Now $F(z)$, though not necessarily $U(z)$, is defined and continuous in the closed disc $|z| \leqslant 1$. Suppose that $F(z) = 1$ for some z in this disc. Then

$$f_\infty + \sum_{n=1}^{\infty} f_n(1 - z^n) = 0,$$

and by considering real parts we see that $f_\infty = 0$ and that

$$f_n(1 - z^n) = 0$$

for all n. Since f_n is non-zero for some n, z is a root of unity. If k is the smallest integer with $z^k = 1$, then $f_n = 0$ unless k divides n. By the remarks following Theorem 1.4, k divides the period of u, which is 1 by hypothesis. Thus $z = 1$.

It follows then that the function

$$\text{Re}\{U(z) - \tfrac{1}{2}\} = \frac{1 - |F(z)|^2}{2|1 - F(z)|^2}$$

admits a continuous extension to

$$\{z; |z| \leqslant 1, z \neq 1\}.$$

Hence

$$g(\theta) = \lim_{r \uparrow 1} g_r(\theta)$$

exists for all $\theta \neq 0$, and there exists, for each $\delta > 0$, a constant M_δ such that

$$|g_r(\theta)| \leqslant M_\delta \qquad (r < 1, \delta < \theta \leqslant \pi).$$

Therefore γ has a density g in $(0, \pi)$, together perhaps with an atom at 0. Writing u_∞ for the value of γ at this atom, (8) takes the required form (7), with

(9) $$g(\theta) = \frac{1 - |F(e^{i\theta})|^2}{\pi|1 - F(e^{i\theta})|^2}. \qquad \blacklozenge$$

Theorem 1.6. (*Erdös–Feller–Pollard*). *If u is a renewal sequence with period d, then*

(10) $$u_\infty = \lim_{n \to \infty} u_{nd}$$

exists. If $f_\infty > 0$, then $u_\infty = 0$. If $f_\infty = 0$, then

(11) $$u_\infty = d \left(\sum_{n=1}^{\infty} nf_n \right)^{-1},$$

where $u_\infty = 0$ if the series diverges.

Proof. If $d = 1$, u admits the representation (7), and (10) is then a consequence of the Riemann–Lebesgue lemma. For general d, this result applies to ^du, and (10) follows. The fact that $u_\infty = 0$ when $f_\infty > 0$ follows from Theorem 1.5, and it is therefore only necessary to consider the case $f_\infty = 0$.

Write

$$F_r = \sum_{s=r+1}^{\infty} f_s,$$

so that (1.14) gives, for $n \geqslant 1$,

$$u_n = \sum_{r=1}^{n} u_{n-r}(F_{r-1} - F_r)$$

$$= \sum_{r=0}^{n-1} u_{(n-1)-r}F_r - \sum_{r=1}^{n} u_{n-r}F_r.$$

Thus

$$\sum_{r=0}^{n} u_{n-r}F_r = \sum_{r=0}^{n-1} u_{(n-1)-r}F_r,$$

showing that

$$\sum_{r=0}^{n} u_{n-r}F_r$$

is independent of n. Putting $n = 0$ shows that

$$\sum_{r=0}^{n} u_{n-r}F_r = 1.$$

Replacing n by nd, and recalling that u. vanishes except at multiples of d, we have

(12) $$\sum_{r=1}^{n} u_{(n-r)d}F_{rd} = 1,$$

and it is simple to check that

$$S = \sum_{r=1}^{\infty} F_{rd} = d^{-1}\sum_{n=1}^{\infty} nf_n \leqslant \infty.$$

If S is finite, the bounded convergence theorem applies to (12) to give (as $n \to \infty$)

$$\sum_{r=1}^{\infty} u_\infty F_{rd} = 1.$$

On the other hand, if S diverges, Fatou's lemma shows that

$$\sum_{r=1}^{\infty} u_{\infty} F_{rd} \leqslant 1,$$

and in either case $u_{\infty} = S^{-1}$.   ◆

1.4 REGENERATIVE PHENOMENA

The results of probability theory fall into two categories. Some are concerned with the finite-dimensional distributions of processes, and relate for example to the behaviour of these distributions in limiting situations. Others assert properties which, with probability one, are enjoyed by the sample functions of the processes. In the first category come the central limit theorem, results on the asymptotic behaviour of Markov transition probabilities, and most of the arguments so far advanced in this chapter. In the second fall, for example, the strong law of large numbers, the law of the iterated logarithm, and results about the continuity of Markov processes.

If our interest were confined to results of the first type, a detailed study of renewal sequences might satisfy us. But results of the second type continually obtrude. For example, in the Markov chain context of §1.1, the question arises whether the number κ of returns to the initial state is finite or infinite. To answer questions like this in their proper setting, a slightly different point of view is necessary.

Suppose then that

$$X = (X_0, X_1, X_2, \ldots)$$

is a Markov chain with $X_0 = a$, and consider the sequence

(1) $\nu_0 = 0, \nu_1, \nu_2, \ldots$

of values of n with $X_n = a$. It is clear that this sequence contains less information than the sequence X. A knowledge of the ν_r tells us, at any time n, whether or not $X_n = a$, but if $X_n \neq a$, it does not tell us where else in the state space X_n is to be found. In effect therefore the states other than a have been lumped together in a single composite state 'not a'.

More formally, let ψ denote the function defined on S by

(2) $\psi(a) = 1, \qquad \psi(i) = 0 \qquad (i \neq a).$

Then

(3) $Z_n = \psi(X_n)$

defines a random sequence Z which takes the value 1 at the times ν_r, and the value 0 elsewhere. The renewal sequence

$$u_n = p_{aa}^{(n)}$$

is determined by Z, since clearly

(4) $$u_n = \mathbf{P}(Z_n = 1).$$

The process Z is not in general Markovian, and the question arises of describing its stochastic structure. It turns out that (always assuming that $X_0 = a$, i.e. that $Z_0 = 1$) this is determined uniquely by the renewal sequence u. This is not obvious, but it is a consequence of the fact that, for any increasing finite sequence of positive integers

$$0 = t_0 < t_1 < t_2 < \ldots < t_n,$$

Z satisfies

$$\mathbf{P}(Z_{t_1} = Z_{t_2} = \ldots = Z_{t_n} = 1) = \mathbf{P}(X_{t_1} = X_{t_2} = \ldots = X_{t_n} = a)$$

$$= \prod_{r=1}^{n} p_{aa}^{(t_r - t_{r-1})} = \prod_{r=1}^{n} u_{t_r - t_{r-1}}.$$

This property is crucial, and it is useful to have a name for it.

Definition. A stochastic process $Z = (Z_n; n \geqslant 1)$, taking the values 0 and 1 is called a discrete time *regenerative phenomenon* if, whenever integers $t_0, t_1, \ldots, t_n$ satisfy

(5) $$0 = t_0 < t_1 < t_2 < \ldots < t_n,$$

then

(6) $$\mathbf{P}(Z_{t_1} = Z_{t_2} = \ldots = Z_{t_n} = 1) = \prod_{r=1}^{n} u_{t_r - t_{r-1}},$$

for some sequence $(u_n; n \geqslant 1)$.

It is useful to extend the sequence (u_n) by setting $u_0 = 1$; then (6) holds trivially even if the strict inequalities in (5) are replaced by weak ones.

Since Z_t takes only the values 0 and 1, (6) can be written in the alternative form

(7) $$\mathbf{E} \prod_{r=1}^{n} Z_{t_r} = \prod_{r=1}^{n} u_{t_r - t_{r-1}}.$$

Now consider an event A of the form

$$A = \{Z_1 = \alpha_1, Z_2 = \alpha_2, \ldots, Z_N = \alpha_N\},$$

where each α_r is either 0 or 1. Then the product

$$\chi = \prod_{r=1}^{N} (-1)^{\alpha_r + 1}(Z_r + \alpha_r - 1)$$

is equal to 1 when A occurs, and 0 otherwise, so that

$$\mathbf{P}(A) = \mathbf{E}(\chi).$$

Now the product χ may be expanded out as a linear combination of products of the form (7), so that (7) determined the value of $\mathbf{E}(\chi)$ and therefore of $\mathbf{P}(A)$. This justifies the assertion that the sequence (u_n) determines the joint distributions of the process Z.

If a sequence (u_n) is chosen arbitrarily, it may or may not turn out to be possible to find a process Z satisfying (6). Indeed, when we use the above argument to compute $\mathbf{P}(A)$, it may turn out, for some choices of the α_r, to be negative, showing that the choice of (u_n) is not an admissible one. What sequences then are admissible?

Theorem 1.7. *There exists a stochastic process* (Z_n) *satisfying* (6) *if and only if* (u_n) *is a renewal sequence.*

Proof. Suppose that (Z_n) exists satisfying (6), and set

$$f_n = \mathbf{P}(Z_1 = Z_2 = \ldots = Z_{n-1} = 0, Z_n = 1).$$

Then $f_n \geqslant 0$ and

$$\sum_{n=1}^{\infty} f_n = \mathbf{P}(Z_n = 1 \text{ for some } n \geqslant 1) \leqslant 1.$$

Now

$$f_n = \mathbf{E}\{(1 - Z_1)(1 - Z_2) \ldots (1 - Z_{n-1})Z_n\},$$

and if expectations are taken of the identity (for $Z_r = 0$ or 1)

$$Z_n = \sum_{r=1}^{n} (1 - Z_1)(1 - Z_2) \ldots (1 - Z_{r-1})Z_r Z_n$$

we obtain

$$u_n = \sum_{r=1}^{n} f_r u_{n-r} \qquad (n \geqslant 1),$$

where $u_0 = 1$. Hence (u_n) is a renewal sequence.

Conversely, if (u_n) is a renewal sequence, there is a Markov chain with

$$u_n = p_{aa}^{(n)},$$

and (3) defines a regenerative phenomena satisfying (6).   ◆

Regenerative phenomena may arise in ways other than the lumping of states in Markov chains. Suppose for instance that (f_n) satisfies (2.1), and that $Y_1, Y_2, \ldots$ are independent random variables with

$$\mathbf{P}(Y_k = n) = f_n \qquad (n = 1, 2, \ldots, \infty).$$

Define

$$S_k = Y_1 + Y_2 + \ldots + Y_k,$$

and let Z_n be the random variable which is 1 if $S_k = n$ for some k, and 0 otherwise. Then it is easy to check that (Z_n) is a regenerative phenomenon for which (u_n) is just the renewal sequence associated with (f_n). One may imagine this process as describing the renewal of a certain component, whence the name 'renewal sequence'. This model is susceptible of considerable elaboration, but the resulting 'renewal theory' is outside the scope of this account (cf. [7], [69]).

1.5 KALUZA SEQUENCES

Although renewal sequences have been recognised in probability theory only in the last twenty years, they had earlier been considered in connection with the theory of continued fractions. The following result is due to Kaluza [26], in honour of whom any sequence satisfying (1) is called a Kaluza sequence.

Theorem 1.8. *Any sequence $(u_n; n \geqslant 0)$ satisfying*

(1) $$0 < u_n \leqslant u_0 = 1, \qquad u_n^2 \leqslant u_{n-1}u_{n+1}$$

for all $n \geqslant 1$ is a renewal sequence.

Proof. Consider a sequence of independent tosses of a coin in which the probability of a head is p. Let m be a positive integer, and define X_n to be the largest integer r in $1 \leqslant r \leqslant m$ for which a tail occurs at the $(n + r)$th toss. If there is no such r, define $X_n = 0$. Then it is easy to see that (X_n) is a Markov chain on the state space $\{0, 1, 2, \ldots, m\}$, its transition probabilities being given by

$$
\begin{aligned}
p_{00} &= p_{r,r-1} = p & (r = 1, 2, \ldots, m), \\
p_{rm} &= 1 - p & (r = 0, 1, 2, \ldots, m), \\
p_{ij} &= 0 & \text{(otherwise)}.
\end{aligned}
$$

For this chain,

$$
\begin{aligned}
p_{00}^{(n)} &= \mathbf{P}(\text{heads at } n + 1, n + 2, \ldots, n + m | \text{heads at } 1, 2, \ldots, m) \\
&= p^{\min(n,m)}.
\end{aligned}
$$

Hence, for any p in $0 \leqslant p \leqslant 1$ and any positive integer m, the sequence $k = k(m, p)$ defined by

$$k_n = p^n \qquad (n \leqslant m)$$
$$= p^m \qquad (n > m)$$

is a renewal sequence.

Taking $f_1 = p, f_n = 0 \ (n \geqslant 2)$, we see also that

$$k_n = p^n$$

defines a renewal sequence $k(\infty, p)$.

If u is any Kaluza sequence, then

$$v_n = u_{n+1}/u_n$$

is non-decreasing, and converges to a limit $v > 0$. Since $v > 1$ would ultimately contradict $u_n \leqslant 1$, we must have $0 < v \leqslant 1$. If

$$p_n = v_{n-1}/v_n,$$

then $0 < p_n \leqslant 1$ and

$$v_n = v \prod_{r=n+1}^{\infty} p_r.$$

Hence

$$u_n = \prod_{j=0}^{n-1} v_j = \prod_{j=0}^{n-1} \left(v \prod_{r=j+1}^{\infty} p_r \right)$$

$$= v^n \prod_{r=1}^{\infty} p_r^{\min(n,r)}.$$

Thus

$$u = k(\infty, v) \prod_{r=1}^{\infty} k(r, p_r),$$

and $u \in \mathscr{R}$ by Theorem 1.2. $\qquad\qquad\qquad\qquad\qquad\qquad \blacklozenge$

It may be noted that the inequality

$$u_n^2 \leqslant u_{n-1} u_{n+1}$$

is equivalent to the assertion that

(2) $$u_{n+1} - 2\lambda u_n + \lambda^2 u_{n-1} \geqslant 0$$

for all $\lambda > 0$. This shows that the class of Kaluza sequences is a closed convex subset of $\mathscr{R}$. As an example of a Kaluza sequence, let

(3) $$u_n = \int_{(0,1]} x^n \nu(dx),$$

where ν is any probability measure on $(0, 1]$. Then

$$u_{n+1} - 2\lambda u_n + \lambda^2 u_{n-1} = \int (x - \lambda)^2 x^{n-1} \nu(dx) \geqslant 0,$$

so that (2) is satisfied and u is a Kaluza sequence. By taking particular forms for ν, large classes of renewal sequences may be generated.

1.6 NOTES

(i) *Simple random walk.* Let X_n be a Markov chain on the positive and negative integers with transition probabilities

$$
\begin{aligned}
p_{ij} &= p & (j - i = 1) \\
&= q = 1 - p & (j - i = -1) \\
&= 0 & \text{(otherwise)},
\end{aligned}
$$

where $0 < p < 1$. The renewal sequence $u_n = p_{00}^{(n)}$ then has the form

$$u_n = \binom{n}{\frac{1}{2}n}(pq)^{\frac{1}{2}n}$$

when n is even, and 0 when n is odd, so that

$$
\begin{aligned}
U(z) &= (1 - 4pqz^2)^{-\frac{1}{2}}, \\
F(z) &= 1 - (1 - 4pqz^2)^{\frac{1}{2}}, \\
f_n &= 2pqn^{-1}u_{n-2}.
\end{aligned}
$$

Moreover,

$$f_\infty = 1 - F(1) = |p - q|,$$

and the chain is transient unless $p = \frac{1}{2}$. The period is 2, and the aperiodic renewal sequence ^2u satisfies (5.1). Is it of the form (5.3)?

(ii) All Kaluza sequences are monotone decreasing, but not all aperiodic renewal sequences are. For example, if

$$f_1 = 1 - \alpha, \quad f_2 = \alpha, \quad f_3 = \quad f_4 = \ldots = 0 \qquad (0 < \alpha < 1),$$

then

$$u_n = \frac{1 - (-\alpha)^{n+1}}{1 + \alpha},$$

which oscillates for all n. (See also [10], [70].)

(iii) *Kolmogorov's theorem.* The Erdös–Feller–Pollard theorem shows that, in a Markov chain, the limits

$$\pi_j = \lim_{n \to \infty} p_{jj}^{(n)}$$

exist, where $n \to \infty$ through multiples of the period d of the state j. A similar result may be deduced for the non-diagonal transition probabilities $p_{ij}^{(n)} (i \neq j)$ by deriving the identity

(1)
$$p_{ij}^{(n)} = \sum_{r=1}^{n} f_{ij}^{(r)} p_{jj}^{(n-r)},$$

where

$$f_{ij}^{(r)} = \mathbf{P}_i(X_1, X_2, \ldots, X_{r-1} \neq j, X_r = j).$$

From this it follows that, unless $p_{ij}^{(n)} = 0$ for all n, or $p_{ji}^{(n)} = 0$ for all n, there is a unique residue class R_{ij} modulo d in which $p_{ij}^{(n)}$ is positive (except perhaps at a finite number of values of n), and that as $n \to \infty$ through R_{ij},

$$p_{ij}^{(n)} \to \pi_j.$$

Hence the Erdös–Feller–Pollard theorem implies the theorem of Kolmogorov [56] that, in an aperiodic chain,

$$\lim_{n \to \infty} p_{ij}^{(n)}$$

exists for all i, j. Historically however, Kolmogorov's theorem came first, and because of Theorem 1.1 it is possible to use it to prove the Erdös–Feller–Pollard theorem.

(iv) One of the original proofs of the Erdös–Feller–Pollard theorem [16] rested upon Wiener's theorem on the reciprocal of an absolutely convergent Fourier series. If u is aperiodic, $f_\infty = 0$ and $\mu = \Sigma n f_n < \infty$, then

$$[1 - F(e^{i\theta})]/[1 - e^{i\theta}]$$

has an absolutely convergent Fourier series

$$\sum_{n=0}^{\infty} F_n e^{in\theta},$$

which vanishes nowhere. Its reciprocal is

$$(1 - e^{i\theta}) U(e^{i\theta}) = \sum_{n=0}^{\infty} (u_n - u_{n-1}) e^{in\theta},$$

which by Wiener's theorem converges absolutely for all θ. In particular,

(2)
$$\sum_{n=1}^{\infty} |u_n - u_{n-1}| < \infty,$$

which implies that u_n converges to a limit. Of course (2) is stronger than the assertion that $\lim u_n$ exists. The proof can be extended to cover the case $\mu = \infty$, but then only yields the weaker assertion. No counter-example is known to the conjecture that (2) holds for all aperiodic renewal sequences.

(v) *Second-order theory.* Suppose that it is possible to choose the initial distribution p_i so that the chain (X_n) is stationary; this will be so if and only if

$$p_j = \sum_{i \in S} p_i p_{ij}$$

for all $j \in S$. Then, for any $a \in S$, the process $Z_n = \psi(X_n)$ defined by (4.3) is stationary, and its autocovariance function is

$$\mathbf{E}(Z_0 Z_n) - \mathbf{E}(Z_0)^2 = p_a(u_n - p_a).$$

If u is aperiodic, $p_a = u_\infty$, so that (3.7) may be used to write the auto-covariance function in the form

$$\int_0^\pi u_\infty g(\theta) \cos n\theta \, d\theta.$$

Hence $u_\infty g(\theta)$ is the spectral density of the process Z. In [59] Loynes reversed this argument, using results from the theory of stationary processes to establish special cases of (3.7) (cf. [50]).

(vi) *Geometric ergodicity.* Suppose that u is aperiodic, and also that there exists $R > 1$ such that $\Sigma f_n R^n < \infty$. Then $F(z)$ is regular in the disc $\{z; |z| < R\}$, and the equation $F(z) = 1$ has only finitely many roots with $|z| \leqslant R - \delta$ for $\delta > 0$. Moreover, there is at most one such root with $|z| \leqslant 1$, namely $z = 1$. Thus there exists $r > 1$ such that $U(z)$ has no singularities in $|z| \leqslant r$ except a pole at $z = 1$, and it follows that

$$u_n = u_\infty + 0(r^{-n});$$

the convergence of u_n to u_∞ is in this case exponentially fast [72], [35].

(vii) On the other hand, by choosing ν suitably in (5.3), we can construct renewal sequences in which the convergence of u_n to u_∞ is arbitrarily slow. For an interesting application to ergodic theory, see [66].

2

(viii) *Removal of geometric factors.* If u is aperiodic, and the limit ρ of Theorem 1.4 satisfies $\rho < 1$, then $u_\infty = 0$. But we may apply Theorem 1.6 to $\bar{u}$, where $\bar{u}_n = u_n \rho^{-n}$, to conclude that

$$\bar{u}_\infty = \lim_{n \to \infty} u_n \rho^{-n}$$

exists. If $\bar{u}_\infty > 0$, then

$$\lim_{n \to \infty} (u_{n+1}/u_n) = \rho.$$

The existence of this limit is closely related to the 'strong ratio limit problem' for Markov chains [6], [53], [62]. A typical result in this area is that, if u is an aperiodic renewal sequence with $u_\infty = f_\infty = 0$, and if

$$\limsup_{n \to \infty} (u_{n+1}/u_n) \leqslant 1,$$

then

$$\lim_{n \to \infty} (u_{n+1}/u_n) = 1.$$

(ix) The class of Kaluza sequences is convex, but $\mathscr{R}$ is not.

(x) *Infinite divisibility.* For any renewal sequence u, define

$$K(u) = \{\alpha > 0; u^\alpha \in \mathscr{R}\},$$

where

$$u^\alpha = (u_0^\alpha, u_1^\alpha, u_2^\alpha, \ldots).$$

By Theorem 1.2, $K(u)$ is an additive semigroup containing the positive integers, and by Theorem 1.3, $K(u)$ is closed in $(0, \infty)$. Very little else is known about $K(u)$. Is it true for instance that, for all $u \in \mathscr{R}$, $K(u) \supseteq [1, \infty)$?

Kendall [32] has considered the possibility that $K(u)$ might have 0 as a limit point, and has shown that this implies that $K(u) = (0, \infty)$. Renewal sequences with this property are said to be *infinitely divisible*. He has shown that a renewal sequence with period d is infinitely divisible if and only if $^d u$ is a Kaluza sequence. In [33] these results are placed in the setting of the beautiful theory of delphic semigroups, which has since been developed further by Davidson [12], [13].

(xi) *Another closure property of $\mathscr{R}$.* If u and v are renewal sequences, construct independent Markov chains X, Y on state spaces S, $\bar{S}$, with transition probabilities p, $\bar{p}$, so that for some $a \in S$, $b \in \bar{S}$,

$$u_n = p_{aa}^{(n)}, \qquad v_n = \bar{p}_{bb}^{(n)}.$$

Repeatedly toss a coin with probability p of falling heads, and denote by $h(n)$ and $t(n)$ the number of heads and tails after n tosses. Then it is not difficult to verify that

$$Z_n = (X_{h(n),}, Y_{t(n)})$$

defines a Markov chain with state space $S \times \bar{S}$. The renewal sequence

$$w_n = \mathbf{P}\{Z_n = (a, b)|Z_0 = (a, b)\}$$

of the state (a, b) is given by

(3)
$$w_n = \sum_{r=0}^{n} \binom{n}{r} p^r (1 - p)^{n-r} u_r v_{n-r}.$$

Hence (3) defines a renewal sequence w for any u, $v \in \mathcal{R}$ and any p in $0 \leqslant p \leqslant 1$. (An unpublished result of R. Davidson.)

(xii) *Bloomfield's inequality* [5]. Let u be an aperiodic renewal sequence, with limit u_∞ as $n \to \infty$. Letting $m \to \infty$ in (3.1), we get the useful inequality

(4)
$$u_n \geqslant 2 - u_\infty^{-1},$$

a result which is of course significant only if $u_\infty > \frac{1}{2}$. Bloomfield has refined this lower bound for u_n by the following argument. Write

$$m_n = \min(u_1, u_2, \ldots, u_n).$$

Then (1.14) implies that, for $n \geqslant 2$,

$$u_n \geqslant \sum_{r=1}^{n} f_r m_{n-1} = m_{n-1}(1 - F_n),$$

where we are assuming that $f_\infty = 0$, so that

$$F_n = \sum_{s=n+1}^{\infty} f_s = 1 - \sum_{r=1}^{n} f_r.$$

Since

$$m_n = \min(u_n, m_{n-1}),$$

we have

$$m_n \geqslant m_{n-1}(1 - F_n),$$

so that

$$m_n \geqslant u_1 \prod_{r=2}^{n} (1 - F_r) = \prod_{r=1}^{n} (1 - F_r),$$

which implies that

$$(5) \qquad u_n \geqslant \prod_{r=1}^{n} (1 - F_r).$$

This is true whenever $f_\infty = 0$, but it is significant only when $u_1 > 0$ (in which case u is automatically aperiodic). Since $0 \leqslant F_n \leqslant F_1 = 1 - u_1$,

$$\log (1 - F_n) \geqslant F_n \log u_1/(1 - u_1),$$

so that

$$(6) \qquad u_n \geqslant \exp \left\{ \frac{\log u_1}{1 - u_1} \sum_{r=1}^{n} F_r \right\}.$$

In particular, since

$$\sum_{r=1}^{n} F_r \leqslant \sum_{r=1}^{\infty} F_r = u_\infty^{-1} - 1,$$

we have

$$(7) \qquad u_n \geqslant \exp \{\log u_1(u_\infty^{-1} - 1)/(1 - u_1)\}.$$

It would be interesting to know how near (7) comes to being a sharp lower bound for $(\min u_n)$ in terms of u_1 and u_∞.

(xiii) Let Z be a regenerative phenomenon with renewal sequence u, and denote by κ the number of values of $n \geqslant 1$ for which $Z_n = 1$. Then if $f_\infty = 0$, $\mathbf{P}(\kappa = \infty) = 1$, but if $f_\infty > 0$, $\mathbf{P}(\kappa < \infty) = 1$, and for all $r \geqslant 0$,

$$\mathbf{P}(\kappa = r) = (1 - f_\infty)f_\infty^r.$$

(xiv) *Forward and backward recurrence times.* If

$$F_n = \min \{r \geqslant 0; Z_{n+r} = 1\},$$
$$B_n = \min \{r \geqslant 0; Z_{n-r} = 1\},$$

then (F_n), (B_n) and (F_n, B_n) are all Markov chains (cf. §4.3), and

$$\{F_n = 0\} = \{B_n = 0\} = \{Z_n = 1\}.$$

This fact furnishes an alternative proof of Chung's Theorem 1.1.

(xv) *The Markov characterisation problem.* Theorem 1.1 is equivalent to the following assertion.

In order that a sequence $(a_n; n \geqslant 0)$ should be expressible in the form $a_n = p_{ii}^{(n)}$ for some state i in some Markov chain, it is necessary and sufficient that (a_n) be a renewal sequence.

The theorem therefore characterises the sequences $(p_{ij}^{(n)}; n \geqslant 0)$ in the diagonal case $i = j$, and the question arises of giving a similar characterisation for the non-diagonal case $i \neq j$. Using (1), the following assertion is easily proved.

In order that a sequence $(b_n; n \geqslant 0)$ *should be expressible in the form* $b_n = p_{ij}^{(n)}$ *for some pair* i, j *of distinct states in some Markov chain, it is necessary and sufficient that*

$$(8) \qquad b_n = \sum_{r=1}^{n} c_r u_{n-r},$$

where $c_r \geqslant 0$,

$$\sum_{r=1}^{\infty} c_r \leqslant 1$$

and (u_n) *is a renewal sequence.*

But this does not really solve the problem, since it is not clear how to decide of a given sequence (b_n) whether it can be written in the form (8). It would be of great interest to have an effective criterion for deciding this question.

(xvi) *Terminology.* The terminology used in the literature has become somewhat confusing. Feller spoke of 'recurrent events', by which he meant collections $\mathscr{E} = \{E(n); n \geqslant 1\}$ of events with the property that the stochastic structure of $\{E(m + n); n \geqslant 1\}$ conditional on $E(m)$ is the same as the structure of $\mathscr{E}$ itself. It is immediate that this is the same as saying that the indicators

$$Z_n = 1 \qquad \text{if } E(n) \text{ occurs,}$$
$$ = 0 \qquad \text{otherwise,}$$

form a regenerative phenomenon in the sense of this chapter.

It clearly makes no essential difference whether one uses the events $E(n)$ or their indicator variables Z_n. In developing the continuous time analogue of Feller's theory I abandoned the word 'recurrent' (which had by then come to be used to express the property $\kappa = \infty$) in favour of the term 'regenerative'. Later Kendall pointed out that it was illogical to call a collection of events an event, and suggested the term 'phenomenon'. Finally, I have come to believe that the indicator process Z is more convenient than the collection $\mathscr{E}$, and it is therefore Z which, in this book, bears the name 'regenerative phenomenon'.

In [6] Chung calls $\mathscr{E}$ a 'repetitive pattern' and gives a careful, if opaque, discussion of its definition.

CHAPTER 2

Regenerative Phenomena in Continuous Time

2.1 MARKOV CHAINS

The account of Feller's theory of recurrent events (discrete time regenerative phenomena) given in Chapter 1 is not, nor is it intended to be, the simplest way of describing these processes. In particular, it deliberately plays down the 'renewal' aspect of these processes, the fact that the instants v_0, v_1, v_2, . . . at which $Z_n = 1$ are the successive partial sums of a sequence of independent random variables with the same distribution.

The reason for this distortion of the natural approach is that, when dealing with continuous time processes, and Markov chains in particular, there is in general no natural analogue of this renewal property. Indeed, the search for renewal processes in Markov chains has been something of a false trail, for reasons which will I hope become apparent.

Let $(X(t); t \geqslant 0)$ be a continuous time Markov chain on the (countable) state space S, that is, a process taking values in S and satisfying the equation

(1) $$\mathbf{P}(X(t) = j \mid X(s) = i, A) = p_{ij}(t - s)$$

for $i, j \in S$, $0 \leqslant s < t$, where A is any event determined by the random variables $X(u)$ $(u < s)$. Here p_{ij} is a function on $(0, \infty)$, and it is immediate [6] that the family $(p_{ij}; i, j \in S)$ satisfies the conditions

(2) $$p_{ij}(t) \geqslant 0, \qquad \sum_{j \in S} p_{ij}(t) = 1,$$

(3) $$p_{ij}(s + t) = \sum_{k \in S} p_{ik}(s) p_{kj}(t).$$

Conversely, if a family of functions p_{ij} satisfies (2) and (3), then there exists a process X satisfying (1), and moreover the distribution of $X(0)$ may be arbitrarily assigned.

Chung [6] has shown that, if the functions p_{ij} are Lebesgue measurable, then there is no loss of generality in assuming that the chain is *standard* in the sense that

$$(4) \qquad \lim_{t \to 0} p_{ij}(t) = \delta_{ij}$$

(where $\delta_{ij} = 1$ if $i = j$ and 0 otherwise), and this assumption will be made without further comment. Then the process X is continuous in probability, so that [15] it may be taken to be both separable and measurable.

Now fix a state $a \in S$, and suppose that $X(0) = a$. Then the set

$$(5) \qquad \{t \geqslant 0;\ X(t) = a\}$$

of time instants at which the process is in its initial state is no longer a sequence of isolated points. Typically it is a set of positive measure, in simple cases a union of intervals.

The renewal theory approach therefore fails in its simple form (but see §2.7 (x)). The discussion of the last chapter, however, generalises without difficulty. Thus consider the process

$$(6) \qquad Z(t) = \psi\{X(t)\},$$

where as before $\psi \colon S \to \{0, 1\}$ satisfies $\psi(a) = 1$, $\psi(i) = 0$ $(i \neq a)$. Then Z has the property that, whenever

$$(7) \qquad 0 = t_0 < t_1 < t_2 < \ldots < t_n,$$
$$\mathbf{P}\{Z(t_1) = Z(t_2) = \ldots = Z(t_n) = 1\}$$
$$= \mathbf{P}_a\{X(t_1) = X(t_2) = \ldots = X(t_n) = a\}$$
$$= \prod_{r=1}^{n} p_{aa}(t_r - t_{r-1}).$$

Exactly as in the discrete-time case, it will be shown that this equation determines the finite-dimensional distributions of Z in terms of the function p_{aa}, and that the existence of Z entails conditions which must be satisfied by the function p_{aa}.

Historically, the reason for the development of the theory was a desire to understand what functions could arise as transition functions p_{ij} in continuous time Markov chains. The study of the process Z seemed, and turned out to be, the natural tool for solving the diagonal case of this characterisation problem. Thus one of the goals of the theory will be the determination of the class, denoted by $\mathscr{PM}$, of diagonal Markov transition functions; a function p belongs to $\mathcal{PM}$ if and only if there is an array of functions p_{ij} satisfying (2), (3) and (4), and such that, for some i,

$$p(t) = p_{ii}(t).$$

2.2   DEFINITIONS AND ELEMENTARY PROPERTIES

As in discrete time, we give a definition which isolates the particular stochastic structure of the process (1.6).

Definition. A stochastic process $Z = (Z(t); t > 0)$, taking the values 0 and 1, is said to be a *regenerative phenomenon* if there exists a function p on $(0, \infty)$ (called the *p-function* of Z) such that, whenever

$$0 = t_0 < t_1 < t_2 < \ldots < t_n,$$

we have

(1) $\mathbf{P}\{Z(t_1) = Z(t_2) = \ldots = Z(t_n) = 1\} = \prod_{r=1}^{n} p(t_r - t_{r-1}).$

The discussion of the previous section shows that the process (1.6) is a regenerative phenomenon with p-function p_{aa}. In view of (1.4), this p-function also satisfies

(2) $\lim_{t \to 0} p(t) = 1.$

A p-function satisfying (2) is said to be *standard*, and the same adjective applies to the regenerative phenomenon. Thus every function in $\mathcal{PM}$ is a standard p-function; if $\mathcal{P}$ denotes the class of standard p-functions then

(3) $\mathcal{PM} \subseteq \mathcal{P}.$

The great difficulty of the theory, which cuts it off from the discrete time theory, is that this inclusion is in fact strict. There are, as will shortly be seen, standard p-functions which cannot arise from Markov chains.

The development of the theory of regenerative phenomena begins with arguments just like those of §1.4. We first rewrite (1) in the form

(4) $\mathbf{E} \prod_{r=1}^{n} Z(t_r) = \prod_{r=1}^{n} p(t_r - t_{r-1}).$

Then, for any values of $t_1, t_2, \ldots, t_N$, and $\alpha_1, \alpha_2, \ldots, \alpha_N \in \{0, 1\}$, we have

(5) $\mathbf{P}\{Z(t_r) = \alpha_r (1 \leqslant r \leqslant N)\} = \mathbf{E} \prod_{r=1}^{N} (-1)^{\alpha_r + 1}(Z_r + \alpha_r - 1).$

The product may be expanded to give a linear combination of products of values of Z, whose expectations are determined by (4). Thus the finite-dimensional distributions of the process Z are expressible in terms of p by equations of the form

$$\text{(6)} \qquad \mathbf{P}\{Z(t_r) = \alpha_r \ (1 \leqslant r \leqslant N)\},$$
$$= \Phi(t_1, t_2, \ldots, t_N; \alpha_1, \alpha_2, \ldots, \alpha_N; p)$$

where, for given α_r, Φ is a polynomial in the numbers $p(t_r - t_s)$ $(0 \leqslant s < r \leqslant N)$.

Fix a positive number T, and let

$$f = \prod_{r=1}^{m} Z(t_r), \qquad g = \prod_{s=1}^{n} Z(u_s),$$

where

$$0 = t_0 < t_1 < \ldots < t_m < T < u_1 < \ldots < u_n.$$

Then (4) shows at once that

$$\text{(7)} \qquad \mathbf{E}\{fZ(T)g\} = \mathbf{E}\{fZ(T)\}\mathbf{E}\{g'\},$$

where

$$g' = \prod_{s=1}^{n} Z(u_s - T)$$

is obtained from g by subtracting T from each time variable u_s. Thus (7) holds also if f and g are linear combinations of products of the above form. In particular,

$$\text{(8)} \quad \mathbf{P}\{Z(t_r) = \alpha_r \ (1 \leqslant r \leqslant m), Z(T) = 1, Z(u_s) = \beta_s \ (1 \leqslant s \leqslant n)\}$$
$$= \mathbf{P}\{Z(t_r) = \alpha_r \ (1 \leqslant r \leqslant m), Z(T) = 1\}$$
$$\times \mathbf{P}\{Z(u_s - T) = \beta_s \ (1 \leqslant s \leqslant n)\}.$$

It then follows from the Kolmogorov extension theorem that, if A is any member of the smallest σ-field with respect to which $Z(t)$ is measurable for all $t < T$, and if B belongs to the smallest σ-field with respect to which $Z(t)$ is measurable for all $t > T$, then

$$\text{(9)} \qquad \mathbf{P}\{A_\cap\{Z(T) = 1\}_\cap B\} = \mathbf{P}\{A_\cap\{Z(T) = 1\}\}\mathbf{P}\{B'\},$$

where B' is obtained from B by the shift $u \to u - T$. This is an important property of conditional independence; the conditional joint distributions of $(Z(T + t); t > 0)$, given that $Z(T) = 1$ and given the past before T, are the same as the unconditional distributions of $(Z(t); t > 0)$. This is loosely expressed by saying that the process Z regenerates at T if $Z(T) = 1$.

The probabilities (6) must, of course, be non-negative, so that any p-function p must satisfy the inequalities

(10) $$\Phi(t_1, t_2, \ldots, t_N; \alpha_1, \alpha_2, \ldots, \alpha_N; p) \geqslant 0.$$

By choosing different values of the α_r this gives an infinite family of inequalities, 2^N for each value of N. Of these 2^N, however, only 2 are not implied by the inequalities for smaller values of N, since by (8) if $\alpha_k = 1$ for some $k \neq N$, then

$$\Phi(t_1, \ldots, t_N; \alpha_1, \ldots, \alpha_N; p)$$
$$= \Phi(t_1, \ldots, t_k; \alpha_1, \ldots, \alpha_k; p)$$
$$\times \Phi(t_{k+1} - t_k, \ldots, t_N - t_k; \alpha_{k+1}, \ldots, \alpha_N; p).$$

The inequality (10) therefore only gives new information if

$$\alpha_1 = \alpha_2 = \ldots = \alpha_{N-1} = 0.$$

Moreover, if we write

$$F_N = F(t_1, \ldots, t_N; p) = \Phi(t_1, \ldots, t_N; 0, \ldots, 0, 1; p),$$

then

$$\Phi(t_1, \ldots, t_N; 0, \ldots, 0, 0; p) = 1 - \sum_{n=1}^{N} F_n,$$

so that the inequalities (10) are equivalent to the inequalities

$$F(t_1, t_2, \ldots, t_N; p) \geqslant 0, \qquad \sum_{n=1}^{N} F(t_1, t_2, \ldots, t_n; p) \leqslant 1.$$

Using (5) we may express the polynomials F explicitly:

(11) $$F(t_1, t_2, \ldots, t_N; p)$$
$$= p(t_N) - \sum_{1 \leqslant i < N} p(t_i)p(t_N - t_i)$$
$$+ \sum_{1 \leqslant i < j < N} \sum p(t_i)p(t_j - t_i)p(t_N - t_j)$$
$$- \ldots + (-1)^{N-1} p(t_1)p(t_2 - t_1) \ldots p(t_N - t_{N-1}).$$

Summing up these results, we have the following theorem.

Theorem 2.1. *For a given function p on $(0, \infty)$, there exists a regenerative phenomenon Z with p-function p if and only if p satisfies, for all $N \geqslant 1$, the inequalities*

(12) $$F(t_1, t_2, \ldots, t_N; p) \geqslant 0, \qquad \sum_{n=1}^{N} F(t_1, t_2, \ldots, t_N; p) \leqslant 1$$

whenever $0 < t_1 < t_2 < \ldots < t_N$ *and F is given by* (11). *The finite-dimensional distributions of the process Z are uniquely determined by the function p. If* $p(T) > 0$ *and E is any event of positive probability in the smallest σ-field with respect to which Z(s) is measurable for all $s < T$, then the process*

$$(13) \qquad ZT(t) = Z(T + t), \qquad (t > 0),$$

under the probability measure $\mathbf{P}(\cdot \mid E, Z(T) = 1)$ *is a regenerative phenomenon with p-function p.*

When $N = 1$, (12) merely asserts that

$$(14) \qquad 0 \leqslant p(t) \leqslant 1.$$

The inequalities with $N = 2$ are less trivial; writing $s = t_1$, $t = t_2 - t_1$ they take the form

$$(15) \qquad p(s)p(t) \leqslant p(s + t) \leqslant p(s)p(t) + 1 - p(s),$$

(cf. (1.3.1)).

Although Theorem 2.1 answers in theory the question of which functions are p-functions, the criteria it suggests are too complicated for many purposes. It would for instance be very hard to use it to prove the following important result.

Theorem 2.2. *If p_1 and p_2 are p-functions, then so is the function*

$$(16) \qquad p(t) = p_1(t)p_2(t).$$

In particular, $\mathscr{P}$ is closed under pointwise multiplication.

Proof. Construct independent regenerative phenomena Z_1 and Z_2 with respective p-functions p_1 and p_2. Then

$$Z(t) = Z_1(t)Z_2(t)$$

satisfies

$$\mathbf{P}\{Z(t_r) = 1 \ (1 \leqslant r \leqslant m)\} = \mathbf{P}\{Z_1(t_r) = Z_2(t_r) = 1 \ (1 \leqslant r \leqslant m)\}$$

$$= \prod_{\alpha=1}^{2} \mathbf{P}\{Z_\alpha(t_r) = 1 \ (1 \leqslant r \leqslant m)\}$$

$$= \prod_{\alpha=1}^{2} \prod_{r=1}^{m} p_\alpha(t_r - t_{r-1})$$

$$= \prod_{r=1}^{m} p(t_r - t_{r-1}),$$

showing that Z is a regenerative phenomenon with p-function p. Moreover, if p_1 and p_2 are standard, then so is p.   ◆

2.3 STANDARD p-FUNCTIONS

For standard p-functions (those satisfying (2.2)) the inequalities (2.15) have a number of simple and useful consequences, both for the p-function itself and for the corresponding regenerative phenomenon. To save complicating the account, we shall always extend the definition of any standard p-function from $(0, \infty)$ to $[0, \infty)$ by setting $p(0) = 1$.

Theorem 2.3. *If p is a standard p-function, then*

(i) $p(t) > 0$ *for all* $t \geqslant 0$,

(ii) p *is uniformly continuous on* $[0, \infty)$,

(iii) *the limit*

(1) $$q = \lim_{t \to 0} t^{-1}\{1 - p(t)\}$$

exists in $0 \leqslant q \leqslant \infty$, *and*

(2) $$p(t) \geqslant e^{-qt},$$

(iv) *if $q < \infty$, p satisfies the Lipschitz condition*

(3) $$|p(t_1) - p(t_2)| \leqslant q|t_1 - t_2|.$$

Proof. Since $p(t) \to 1$ as $t \to 0$, $p(t) > 0$ for all $t < t_0$ for some $t_0 > 0$. The left hand inequality of (2.15) shows that, for any $n \geqslant 1$,

$$p(t) \geqslant p(t/n)^n,$$

and taking $n > t/t_0$ shows that $p(t) > 0$.

From (2.15),

$$-p(t)\{1 - p(s)\} \leqslant p(s + t) - p(t) \leqslant \{1 - p(s)\}\{1 - p(t)\},$$

so that

$$|p(s + t) - p(t)| \leqslant 1 - p(s),$$

i.e.

(4) $$|p(t_1) - p(t_2)| \leqslant 1 - p(|t_1 - t_2|),$$

and (2.2) shows that p is uniformly continuous.

Since $0 < p(t) \leqslant 1$, the function

$$\phi(t) = - \log p(t)$$

is finite and non-negative, and (2.15) shows that

$$\phi(s + t) \leqslant \phi(s) + \phi(t).$$

A famous result on subadditive functions (Theorem 7.11.1 of [24]) asserts that, if

(5) $$q = \sup_{t>0} \phi(t)/t,$$

then

$$q = \lim_{t \to 0} \phi(t)/t.$$

Since $p(t) \to 1$, $\phi(t) \to 0$, so that

$$\frac{1 - p(t)}{t} = \frac{1 - e^{-\phi(t)}}{t} \sim \frac{\phi(t)}{t},$$

and (1) is proved; (2) then follows from (5). Finally, if $q < \infty$, (2) and (4) give

$$|p(t_1) - p(t_2)| \leqslant 1 - e^{-q|t_1 - t_2|} \leqslant q|t_1 - t_2|. \qquad \blacklozenge$$

Theorem 2.4. *A standard regenerative phenomenon is continuous in probability, and indeed*

(6) $$\mathbf{P}\{Z(t_1) \neq Z(t_2)\} \leqslant 2\{1 - p(|t_1 - t_2|)\}.$$

Proof. Suppose without loss of generality that $t_1 < t_2$; then

$$\begin{aligned}
\mathbf{P}\{Z(t_1) \neq Z(t_2)\} &= \mathbf{P}\{Z(t_1) = 0, Z(t_2) = 1\} + \mathbf{P}\{Z(t_1) = 1, Z(t_2) = 0\} \\
&= \{p(t_2) - p(t_1)p(t_2 - t_1)\} + p(t_1)\{1 - p(t_2 - t_1)\} \\
&\leqslant \{1 - p(t_2 - t_1)\} + p(t_1)\{1 - p(t_2 - t_1)\} \\
&\leqslant 2\{1 - p(t_2 - t_1)\}.
\end{aligned}$$

This tends to 0 as $t_2 - t_1 \to 0$. $\qquad \blacklozenge$

The importance of this result is of course that it brings into play the powerful theory of Doob [15], according to which versions of Z may be chosen with desirable properties such as measurability and separability. The question of choice of versions will be taken up again in Chapter 4.

It is appropriate at this stage to exhibit a large class of p-functions with a rather explicit representation. Suppose that u is any renewal sequence. Then by Theorem 1.1 there exists a discrete time Markov chain with

$$u_n = p_{aa}^{(n)}$$

for some state a. From this a continuous time Markov chain may be constructed by letting the transitions of the former chain take place at the instants of a Poisson process of constant rate λ (say). More formally, the functions

(7) $$p_{ij}(t) = \sum_{n=0}^{\infty} p_{ij}^{(n)} \pi_n(\lambda t),$$

where $p_{ij}^{(0)} = \delta_{ij}$ and

(8) $$\pi_n(\mu) = e^{-\mu} \mu^n / n!$$

is the usual Poisson probability, satisfy (1.2), (1.3) and (1.4), so that

$$p_{aa} \in \mathscr{P}\mathscr{M} \subseteq \mathscr{P}.$$

Since

$$p_{aa}(t) = \sum_{n=0}^{\infty} u_n \pi_n(\lambda t),$$

we have proved that *every function of the form*

(9) $$p(t) = \sum_{n=0}^{\infty} u_n \pi_n(\lambda t),$$

for $\lambda > 0$ and $u \in \mathscr{R}$, belongs to $\mathscr{P}\mathscr{M}$ and hence to $\mathscr{P}$.

If we write $\mathscr{Q}$ for the class of functions expressible in the form (9), then

(10) $$\mathscr{Q} \subset \mathscr{P}\mathscr{M} \subseteq \mathscr{P},$$

the strictness of the first inclusion following from the fact that every member of $\mathscr{Q}$ has

$$q = \lambda(1 - u_1) < \infty,$$

whereas [6] many members of $\mathscr{P}\mathscr{M}$ have $q = \infty$.

In a sense, however, $\mathscr{Q}$ is quite large. We shall see in the next section that (in the topology of pointwise convergence) $\mathscr{Q}$ is dense in $\mathscr{P}$. It will become apparent that $\mathscr{Q}$ and $\mathscr{P}$ are fairly easy to handle, and that the difficult problem is to locate $\mathscr{P}\mathscr{M}$ between them.

2.4. THE METHOD OF SKELETONS

Let Z be any regenerative phenomenon with p-function p, and fix any positive number h. In the defining equation (2.1) put $t_r = n_r h$, where the n_r are integers. The resulting equation shows that the sequence

(1) $$^h Z = (Z(nh); n = 1, 2, \ldots)$$

is a discrete time regenerative phenomenon, the discrete skeleton of Z at scale h. The renewal sequence u of ${}^hZ$ is given by

$$u_n = \mathbf{P}\{Z(nh) = 1\} = p(nh).$$

Thus we have the following result.

Theorem 2.5. *If p is any p-function, and h any positive number, then*

$$(2) \qquad\qquad (1, p(h), p(2h), p(3h), \ldots)$$

is a renewal sequence.

It is rather easy to check that the sequence (f_n) corresponding to the renewal sequence (2) is given, in the notation of §2.2, by

$$(3) \qquad\qquad f_n = f_n(h) = F(h, 2h, \ldots, nh; p).$$

The converse to Theorem 2.5 is false, but there is a partial converse (§3.6 (iv)).

Consequences of Theorem 2.5 are legion, and the most important one is deferred to the next chapter, but simple applications include the following.

Theorem 2.6. *If p is any standard p-function, then*

$$(4) \qquad\qquad p(\infty) = \lim_{t \to \infty} p(t)$$

exists.

Proof. By Theorem 2.3(i) the renewal sequence (2) is aperiodic, and by Theorem 1.6,

$$l(h) = \lim_{n \to \infty} p(nh)$$

exists for all $h > 0$. Write

$$\psi(h) = \sup_{0 < t \leqslant h} \{1 - p(t)\},$$

so that (3.4) implies that, for any t,

$$|p(t) - p([t/h]h)| \leqslant \psi(h).$$

Thus

$$p(t) \leqslant p([t/h]h) + \psi(h),$$

whence

$$\limsup_{t \to \infty} p(t) \leqslant l(h) + \psi(h).$$

Similarly

$$\liminf_{t \to \infty} p(t) \geqslant l(h) - \psi(h),$$

so that

$$\limsup_{t \to \infty} p(t) - \liminf_{t \to \infty} p(t) \leqslant 2\psi(h),$$

and this may be made arbitrarily small by taking h small. ◆

Theorem 2.7. *If p is any standard p-function, there exists a sequence (p_k) of functions in $\mathcal{Q}$ such that, for all $t \geqslant 0$,*

$$\lim_{k \to \infty} p_k(t) = p(t).$$

Proof. The sequence $(p(nk^{-1}); n \geqslant 0)$ is a renewal sequence, so that

$$p_k(t) = \sum_{n=0}^{\infty} p(nk^{-1}) \pi_n(kt)$$

belongs to $\mathcal{Q}$. For any $\epsilon > 0$, choose δ so that

$$|p(s) - p(t)| < \tfrac{1}{2}\epsilon \quad \text{for} \quad |s - t| < \delta.$$

Then

$$
\begin{aligned}
|p(t) - p_k(t)| &\leqslant \sum_{n=0}^{\infty} |p(t) - p(nk^{-1})| \pi_n(kt) \\
&< \sum_{|n-kt|<k\delta} \tfrac{1}{2}\epsilon \pi_n(kt) + \sum_{|n-kt|\geqslant k\delta} \pi_n(kt) \\
&< \tfrac{1}{2}\epsilon + t/k\delta^2,
\end{aligned}
$$

using Tchebychev's inequality. Hence

$$|p(t) - p_k(t)| < \epsilon$$

for $k > 2t/\epsilon\delta^2$. ◆

It is useful to express this result in rather more grandiloquent terms. Since p-functions are functions from $(0, \infty)$ into $[0, 1]$, they can be regarded as elements of the cartesian product space $[0, 1]^{(0, \infty)}$, whose product topology is compact and Hausdorff. Since the set of p-functions is given by

$$(5) \quad \bigcap_{N=1}^{\infty} \bigcap_{t_1 < t_2 < \ldots < t_N} \{p \in [0, 1]^{(0, \infty)};$$

$$F(t_1, \ldots, t_N; p) \geqslant 0, \sum_{n=1}^{N} F(t_1, \ldots, t_n; p) \leqslant 1\},$$

where the F are polynomials in the values of p, it is closed and therefore compact. The subset $\mathcal{P}$ of standard p-functions is not compact (consider the sequence (p_n), where $p_n(t) = e^{-nt}$). Theorem 2.7 asserts that $\mathcal{Q}$, and thus *a fortiori* $\mathcal{PM}$, is dense in $\mathcal{P}$.

2.5 EXAMPLES OF REGENERATIVE PHENOMENA

Regenerative phenomena often arise directly, rather than from Markov chains. Consider, for example, a model in which independent positive (possibly infinite) variables $X_1, X_2, \ldots$ satisfy

$$\mathbf{P}(X_n \leqslant x) = A(x) \qquad (n \text{ odd})$$
$$= B(x) \qquad (n \text{ even}).$$

If $S_0 = 0$, $\qquad S_n = X_1 + X_2 + \ldots + X_n \qquad (n \geqslant 1)$,

define a process $(Z(t); t > 0)$ by

$$Z(t) = 1 \qquad (S_{2n} < t \leqslant S_{2n+1})$$
$$= 0 \qquad (S_{2n+1} < t \leqslant S_{2n+2})$$

$(n = 0, 1, 2, \ldots)$. Such a process, which takes the values 1 and 0 alternately on intervals of random length, is called an alternating renewal process [7].

If we write

$$p(t) = \mathbf{P}\{Z(t) = 1\},$$

then

$$p(t) = \sum_{n=0}^{\infty} \mathbf{P}\{S_{2n} < t \leqslant S_{2n+1}\}$$

$$= \sum_{n=0}^{\infty} \{\mathbf{P}(S_{2n} < t) - \mathbf{P}(S_{2n+1} < t)\},$$

so that

(1) $$p(t) = \sum_{n=0}^{\infty} (-1)^n \mathbf{P}(S_n < t),$$

and p is computable in terms of convolutions of A and B. This is conveniently done by Laplace transforms; for $\theta > 0$

$$\int_0^{\infty} p(t) \, e^{-\theta t} \, dt = \int_0^{\infty} \mathbf{E}\{Z(t)\} \, e^{-\theta t} \, dt$$

$$= \mathbf{E} \int_0^{\infty} Z(t) \, e^{-\theta t} \, dt$$

$$= \mathbf{E} \sum_{n=0}^{\infty} \int_{S_{2n}}^{S_{2n+1}} e^{-\theta t} \, dt$$

$$= \theta^{-1} \sum_{n=0}^{\infty} \{\mathbf{E}(e^{-\theta S_{2n}}) - \mathbf{E}(e^{-\theta S_{2n+1}})\}$$

$$= \theta^{-1} \sum_{n=0}^{\infty} \{\alpha(\theta)^n \beta(\theta)^n - \alpha(\theta)^{n+1} \beta(\theta)^n\},$$

where

$$\alpha(\theta) = \int_0^\infty e^{-\theta x} \, dA(x),$$

$$\beta(\theta) = \int_0^\infty e^{-\theta x} \, dB(x).$$

Thus

(2)
$$\int_0^\infty p(t) \, e^{-\theta t} \, dt = \theta^{-1} \frac{1 - \alpha(\theta)}{1 - \alpha(\theta)\beta(\theta)}.$$

In general, Z is not a regenerative phenomenon, but it is in the very special case in which the intervals with $Z = 1$ have a negative exponential distribution

(3)
$$A(x) = 1 - e^{-ax} \quad (x \geqslant 0).$$

For suppose that, for some $T > 0$, $Z(T) = 1$, and let N be the integer with $S_{2N} < T < S_{2N+1}$. Then the basic property of the exponential distribution [19] implies that $S_{2N+1} - T$ has the same distribution (3), and so the process $(Z(T + t); t > 0)$ has the same structure at Z, independently of $(Z(s); s < T)$.

Thus in this case Z is a regenerative phenomenon with p-function p. Since

$$p(t) \geqslant \mathbf{P}(S_1 \geqslant t) = e^{-at},$$

p is standard (indeed $q < \infty$), and therefore continuous and uniquely determined by its Laplace transform. Since

$$\alpha(\theta) = a(a + \theta)^{-1},$$

(2) takes the form

(4)
$$\int_0^\infty p(t) \, e^{-\theta t} \, dt = \{a + \theta - a\beta(\theta)\}^{-1}.$$

The right hand side of (4) may be expanded in the series

$$\sum_{n=0}^\infty \frac{a^n \beta(\theta)^n}{(a + \theta)^{n+1}},$$

which is the Laplace transform of

(5)
$$p(t) = \sum_{n=0}^\infty \int_0^t \pi_n\{a(t - u)\} \, dB_n(u),$$

where B_n is the n-fold Stieltjes convolution of B with itself.

By varying B, we can construct many p-functions, some explicitly. In the next chapter, it will appear that every p-function with $q < \infty$ may be expressed in this form, and that (4) (though not (5)) has a simple generalisation which allows even the case $q = \infty$ to be covered.

Meanwhile, let us examine the case in which B is degenerate, at the point τ, say. Then (5) becomes

$$(6) \qquad p(t) = \sum_{n=0}^{[t/\tau]} \pi_n\{a(t - n\tau)\}.$$

This is an oscillating function, differentiable everywhere except at the point $t = \tau$, where it has right and left derivatives

$$(7) \qquad D_+ p(\tau) = a - a\,e^{-a\tau}, \qquad D_- p(\tau) = -a\,e^{-a\tau}.$$

It is an interesting analytical exercise (made redundant by Theorem 3.3) to show that, as $t \to \infty$, $p(t)$ tends to the limit

$$(8) \qquad p(\infty) = (1 + a\tau)^{-1}.$$

A famous theorem of Ornstein [63] asserts that the transition functions p_{ij} in a Markov chain are continuously differentiable in $(0, \infty)$. Accordingly, (7) debars the function (6) from being of the form $p_{aa}(t)$ in any chain, so that (6) belongs to $\mathscr{P}$ but not to $\mathscr{PM}$. Thus (2.3) is strengthened to the strict inclusion

$$(9) \qquad \mathscr{PM} \subset \mathscr{P}.$$

An interesting example of a p-function which is not standard can be realised (as was noted by Davidson [12]) by setting $\tau = 1 - \lambda a^{-1}$ in (6), and letting $a \to \infty$. The limiting function (which is necessarily a p-function since the space of p-functions is compact) is given by

$$(10) \qquad \begin{aligned} p(t) &= \pi_t(\lambda t) \qquad (t = 0, 1, 2, \ldots) \\ &= 0 \qquad\qquad \text{(otherwise)}. \end{aligned}$$

From Theorem 2.5 we deduce the by no means obvious fact that, for any $\lambda > 0$,

$$(11) \qquad u_n = \pi_n(\lambda n)$$

is a renewal sequence (cf. §2.7(xv)). The general theory of non-standard p-functions will be described in §3.4.

Another example comes from a continuous time version of the argument used in §1.5. Consider a Poisson process

$$0 < X_1 < X_2 < \ldots$$

of rate λ on $(0, \infty)$, and let $Y_1, Y_2, \ldots$ be independent positive random variables with the same distribution function F. Define

$$Z(t) = 0 \quad \text{(if } X_n < t < X_n + Y_n \text{ for some } n)$$
$$= 1 \quad \text{(otherwise).}^*$$

It is not difficult to see that Z is a regenerative phenomenon. Its p-function may be calculated by noting that, conditional on the number N of points X_n in $(0, t)$, the positions of these points have the same joint distribution as those of a sample $X_1', X_2', \ldots, X_N'$ drawn from a uniform distribution on $(0, t)$ and arranged in order. Thus

$$p(t) = \mathbf{E}\{\mathbf{P}[X_n + Y_n < t \, (1 \leqslant n \leqslant N)|N]\}$$
$$= \mathbf{E}\{\mathbf{P}[X_n' + Y_n < t \, (1 \leqslant n \leqslant N)]\}$$
$$= \mathbf{E}\left\{\left[t^{-1} \int_0^t F(t - u) \, du\right]^N\right\}$$
$$= \exp\left\{\lambda t\left[t^{-1} \int_0^t F(t - u) \, du - 1\right]\right\}$$
$$= \exp\left\{-\lambda \int_0^t [1 - F(v)] \, dv\right\}$$
$$(12) \qquad = \exp\left\{-\lambda \int_0^\infty \min{(t, x)} \, dF(x)\right\}.$$

Since λ and F are arbitrary, this means that every function of the form

$$(13) \qquad p(t) = \exp\left\{-\int_{(0, \infty]} \min{(t, x)}\nu(dx)\right\},$$

where ν is a totally finite measure† on $(0, \infty]$, belongs to $\mathscr{P}$. Now the expression (13) makes sense, and defines a function $p(t)$ with

$$\lim_{t \to 0} p(t) = 1,$$

even if ν is not totally finite, so long as the weaker condition

$$(14) \qquad \int_{(0, \infty]} \min{(1, x)}\nu(dx) < \infty$$

---

* This model arises in a number of different contexts, and has been analysed by a variety of techniques; see [51] for a detailed account.
† All measures on the real line are positive Borel measures.

is satisfied. Even in that case, p is a limit of p-functions, since

$$p(t) = \lim_{\epsilon \to 0} \exp \left\{ - \int_{(\epsilon, \infty]} \min (t, x) \nu(dx) \right\}$$

and (14) implies that $\nu(\epsilon, \infty] < \infty$, so that (13) still defines an element of $\mathscr{P}$.

It is clear that, when (14) holds, (13) defines a function with the property that

$$\phi(t) = - \log p(t)$$

is non-negative, continuous and concave, with $\phi(0) = 1$. Conversely, if ϕ has these properties, it has [23] a right derivative ψ which is non-increasing, and so defines a measure ν on $(0, \infty]$ by

$$\nu(x, \infty] = \psi(x).$$

Then

$$\phi(t) = \int_0^t \psi(x) \, dx = \int_{(0, \infty]} \min (t, x) \nu(dx),$$

so that $p(t) = e^{-\phi(t)}$ is of the form (13). Thus $\mathscr{P}$ contains every function of the form

(15) $$p(t) = e^{-\phi(t)},$$

where ϕ is non-negative, continuous and concave on $[0, \infty)$, with $\phi(0) = 0$.

2.6 DELAYED REGENERATIVE PHENOMENA

Let $X(t)$ be a Markov chain, select a state a, and define as usual

$$Z(t) = \psi\{X(t)\},$$

where $\psi(a) = 1$ and $\psi(i) = 0$ for all $i \neq a$. We have seen that, under the probability measure $\mathbf{P}_a$, Z is a regenerative phenomenon with p-function p_{aa}, but what of its structure under the probability measure $\mathbf{P}_b$ (i.e. under the initial condition $X(0) = b$)? The answer is contained in the fact that, for $0 < t_1 < t_2 < \ldots < t_n$,

$$\mathbf{P}_b\{Z(t_1) = Z(t_2) = \ldots = Z(t_n) = 1\}$$
$$= \mathbf{P}_b\{X(t_1) = X(t_2) = \ldots = X(t_n) = a\}$$
$$= p_{ba}(t_1) p_{aa}(t_2 - t_1) \ldots p_{aa}(t_n - t_{n-1}).$$

With this in mind we make the following definition.

Definition. A process $(Z(t); t > 0)$ taking the values 0 and 1 is said to be a *delayed regenerative phenomenon* if there exist functions p, p^0 on $(0, \infty)$ such that, for $0 < t_1 < t_2 < \ldots < t_n$,

$$(1) \qquad \mathbf{P}\{Z(t_r) = 1 \ (1 \leqslant r \leqslant n)\} = p^0(t_1) \prod_{r=2}^{n} p(t_r - t_{r-1}).$$

Exactly as in §2.2, the functions p and p^0 determine the finite-dimensional distributions of Z, and one may ask which pairs (p, p^0) are admissible. The answer is given by the following theorem [39], in which the trivial case in which p^0 vanishes identically is excluded without comment.

Theorem 2.8. *There exists a process Z satisfying (1) if and only if p is a p-function and p^0 belongs to a certain non-empty closed convex set $\mathcal{L}(p)$ depending on p. If p is standard, $\mathcal{L}(p)$ consists exactly of those functions which satisfy*

$$(2) \qquad p^0(t) = \int_{[0,t)} p(t - s)\phi(ds) + \theta_t\phi\{t\},$$

where ϕ is a probability measure on $[0, \infty]$ and $0 \leqslant \theta_t \leqslant 1$ for each atom t of ϕ.

Proof. Choose T so that $p^0(T) > 0$; then (1) shows that, under the conditional probability measure $\mathbf{P}\{\cdot | Z(T) = 1\}$, the process $(Z(T + t); t > 0)$ is a regenerative phenomenon with p-function p. Thus p must be a p-function.

For any p-function p, define $\mathcal{L}(p)$ to be the class of functions p^0 for which there is a process Z satisfying (1); $\mathcal{L}(p)$ is non-empty since it contains p itself. As in §2.2, the process will exist if and only if the probabilities

$$(3) \qquad \mathbf{P}\{Z(t_r) = \alpha_r \ (1 \leqslant r \leqslant n)\}$$

calculated from (1) are non-negative. Now (1) is linear in p^0, and hence (3) is also; in fact it must take the form

$$c_0 + \sum_{r=1}^{n} c_r p^0(t_r),$$

where the coefficients c_r depend on p and on the α_r. Thus $\mathcal{L}(p)$ is the set of functions $p^0 \colon (0, \infty) \to [0, 1]$ satisfying a family of inequalities of the form

$$c_0 + \sum_{r=1}^{n} c_r p^0(t_r) \geqslant 0;$$

it is therefore convex, and closed in the product topology of $[0, 1]^{(0, \infty)}$.

Now suppose that p is standard, that $p^0 \in \mathcal{L}(p)$, and that Z satisfies (1). For positive h, write

$$f_n(h) = \mathbf{P}\{Z(rh) = 0 \; (1 \leqslant r \leqslant n - 1), Z(nh) = 1\},$$

so that $f_n(h) \geqslant 0$ and

$$\sum_{n=1}^{\infty} f_n(h) = \mathbf{P}\{Z(nh) = 1 \text{ for some } n \geqslant 1\} \leqslant 1.$$

Then, for $n \geqslant 1$,

$$p^0(nh) = \sum_{r=1}^{n} \mathbf{P}\{Z(sh) = 0 \; (1 \leqslant s \leqslant r - 1), Z(rh) = Z(nh) = 1\}$$

$$= \sum_{r=1}^{n} f_r(h)p(nh - rh).$$

Thus

$$p^0(nh) = \int_{[0,nh]} p(nh - s)\phi_h(ds),$$

where ϕ_h is the probability measure on $[0, \infty]$ with atoms

$$\phi_h\{nh\} = f_n(h) \; (n = 1, 2, \ldots),$$

$$\phi_h\{\infty\} = 1 - \sum_{n=1}^{\infty} f_n(h).$$

We pause to note that, for any $p^0 \in \mathcal{L}(p)$,

$$\mathbf{P}\{Z(u) = Z(u + v) = 1\} \leqslant \mathbf{P}\{Z(u + v) = 1\},$$

so that

(4) $$p^0(u)p(v) \leqslant p^0(u + v).$$

Now fix $t > 0$, and write

$$n = [t/h], \qquad u = t - nh, \qquad v = 1 - u.$$

Then (4) gives

$$p^0(nh)p(u) \leqslant p^0(t) \leqslant p^0((n + 1)h)/p(v),$$

so that

$$p(u) \int_{[0,nh]} p(nh - s)\phi_h(ds) \leqslant p^0(t)$$

$$\leqslant p(v)^{-1} \int_{[0,(n+1)h]} p((n + 1)h - s)\phi_h(ds).$$

The results of the Appendix allow one to let $h \to 0$ along a sequence such that ϕ_h converges weakly to a probability measure ϕ on $[0, \infty]$, to give

$$\int_{[0,t)} p(t - s)\phi(ds) \leqslant p^0(t) \leqslant \int_{[0,t]} p(t - s)\phi(ds).$$

The extreme terms in this chain of inequalities are equal unless t is an atom of ϕ, in which case they differ by $\phi\{t\}$. Hence (2) is satisfied.

Conversely, suppose that $p \in \mathscr{P}$ and p^0 is given by (2). Let $\tilde{Z}$ be a regenerative phenomenon with p-function p, and T a random variable with distribution ϕ, independent of $\tilde{Z}$. For every atom t of ϕ, let $E(t)$ be an event with probability θ_t, such that the events $E(t)$ are independent of one another and of $\tilde{Z}$, T. Define Z by

$$Z(t) = \tilde{Z}(t - T) \qquad \text{if } T < t,$$
$$= 1 \qquad\qquad \text{if } T = t \text{ and } E(t) \text{ occurs,}$$
$$= 0 \qquad\qquad \text{otherwise.}$$

Then it is trivial to check that (1) is satisfied, so that $p^0 \in \mathscr{L}(p)$.   ◆

The construction given in the last part of the proof explains the reason for the name; the process Z is the regenerative phenomenon $\tilde{Z}$, delayed by a random time T. From (2) follows, by straightforward analytic arguments which will not be detailed, some simple properties of p^0.

Corollary 1. *If $p \in \mathscr{P}$, $p^0 \in \mathscr{L}(p)$, then p^0 is continuous except possibly at a countable number of upward jump discontinuities (the atoms of ϕ). The limits*

$$\lim_{t \to 0} p^0(t) = \phi\{0\}$$

and

$$\lim_{t \to \infty} p^0(t) = p(\infty)\phi[0, \infty)$$

exist. In every finite interval, p^0 has bounded variation.

Let Z be a standard regenerative phenomenon with p-function p. Then, for any positive a, the process

$$\bar{Z}(t) = Z(t + a)$$

satisfies (1), with

$$\bar{p}(t) = p(t), \qquad \bar{p}^0(t) = p(t + a).$$

Hence for any $p \in \mathscr{P}$,

(5) $$p(\cdot + a) \in \mathscr{L}(p).$$

Using the continuity of p, Theorem 2.8 implies the following result.

Corollary 2. *For any $p \in \mathscr{P}$ and $a > 0$, there exists a probability measure $\phi(\cdot, a)$ on $[0, \infty]$, with no atoms in $(0, \infty)$, such that*

$$(6) \qquad p(t + a) = \int_0^t p(t - s)\phi(ds, a).$$

As $a \to \infty$, the function (5) converges to the function which takes the constant value $p(\infty)$. Thus $\mathscr{L}(p)$ contains this constant function, so that there exists a process Z satisfying (for $0 < t_1 < t_2 < \ldots < t_n$)

$$(7) \qquad \mathbf{P}\{Z(t_r) = 1 \ (1 \leqslant r \leqslant n)\} = p(\infty) \prod_{r=2}^n p(t_r - t_{r-1}).$$

Since the right hand side depends only on the differences $t_r - t_{r-1}$, Z is stationary, and is called an *equilibrium regenerative phenomenon*. It is nontrivial only when $p(\infty) > 0$. By a theorem of Doob [15], Z may be extended to a stationary process $(Z(t); -\infty < t < \infty)$, and then (7) holds whenever $t_1 < t_2 < \ldots < t_n$.

The autocovariance function of Z is

$$\mathbf{E}\{Z(0)Z(t)\} - \mathbf{E}\{Z(0)\}\mathbf{E}\{Z(t)\} = p(\infty)p(t) - p(\infty)^2$$

for $t \geqslant 0$, so that if μ is the spectral measure of Z,

$$p(\infty)\{p(t) - p(\infty)\} = \int_0^\infty \cos \omega t \, \mu(d\omega).$$

This shows that, if $p(\infty) > 0$, p admits the Fourier representation

$$(8) \qquad p(t) = p(\infty) + \int_0^\infty \cos \omega t \, \gamma(d\omega),$$

where $\gamma = p(\infty)^{-1}\mu$.

This argument breaks down if $p(\infty) = 0$, or if p is not standard. It was, however, proved in [50] that an equilibrium regenerative phenomenon exists for every p-function, so long as the 'probability' measure $\mathbf{P}$ is not required to be finite.

Theorem 2.9. *If p is any p-function, there exists a measure space $(\Omega, \mathfrak{F}, \mathbf{P})$ and measurable functions $Z(t) \colon \Omega \to \{0, 1\} \ (-\infty < t < \infty)$ with the property that, if $t_1 < t_2 < \ldots < t_n$, then*

$$(9) \qquad \mathbf{P}\{Z(t_1) = Z(t_2) = \ldots = Z(t_n) = 1\} = \prod_{r=2}^n p(t_r - t_{r-1}).$$

If p is standard, then $\mathbf{P}$ is σ-finite, and

$$(10) \qquad \mathbf{P}(\Omega) = p(\infty)^{-1}.$$

Since this theorem will not be required in the sequel, the proof is omitted, and may be found in [50]. Notice that, for any T, the restriction of $\mathbf{P}$ to $\{Z(T) = 1\}$ is a probability measure, with respect to which $(Z(T + t);\ t > 0)$ and $(Z(T - t);\ t > 0)$ are independent regenerative phenomena with p-function p.

Equation (9) shows that, for any real $c_1, c_2, \ldots, c_n$,

$$0 \leqslant \int_\Omega \left\{ \sum_{r=1}^n c_r Z(t_r) \right\}^2 d\mathbf{P}$$

$$= \sum_{r,s=1}^n c_r c_s \int_\Omega Z(t_r) Z(t_s)\, d\mathbf{P}$$

$$= \sum_{r,s=1}^n c_r c_s \mathbf{P}\{Z(t_r) = Z(t_s) = 1\}$$

$$= \sum_{r,s=1}^n c_r c_s p(|t_r - t_s|).$$

In other words, every p-function, standard or not, is positive-definite. If $p \in \mathscr{P}$, p is continuous, and Bochner's theorem implies a Fourier representation of the form (8).

Now set $t = nh$, and compare the resulting representation for the renewal sequence $(p(nh))$:

$$p(nh) = p(\infty) + \int_0^\infty \cos(\omega nh) \gamma(d\omega)$$

$$= p(\infty) + \int_0^{2\pi/h} \cos(\omega nh) \sum_{m=0}^\infty \gamma(2m\pi h^{-1} + d\omega)$$

with (1.3.7). This shows that γ is absolutely continuous, so that (8) takes the form

$$(11) \qquad p(t) = p(\infty) + \int_0^\infty \cos \omega t\, g(\omega)\, d\omega$$

for a non-negative function g. A quite different proof of this result, using the theorems to be established in the next chapter, will be found in [38].

2.7 NOTES

(i) *History.* The basic results of this chapter were first published in [37], [38] and [39], but many of the arguments were directly inspired by those that had earlier been used for the special case of a Markov chain. As far as I know, the first suggestion of a general theory came from Bartlett [1], to whom equation (5.4) is due.

(ii) *The basic inequalities.* The inequalities (2.12) which define a p-function may be thrown into determinantal form; p is a p-function if and only if, for all $n \geqslant 1$ and all $0 < t_1 < t_2 < \ldots < t_n$,

$$
\begin{vmatrix}
1 & p(t_1) & p(t_2) & \cdots & p(t_n) \\
1 & 1 & p(t_2 - t_1) & \cdots & p(t_n - t_1) \\
1 & 0 & 1 & \cdots & p(t_n - t_2) \\
\cdots & \cdots & \cdots & \cdots & \cdots \\
1 & 0 & 0 & \cdots & 1
\end{vmatrix} \geqslant 0
$$

and

$$
(-1)^{n-1}
\begin{vmatrix}
p(t_1) & p(t_2) & p(t_3) & \cdots & p(t_n) \\
1 & p(t_2 - t_1) & p(t_3 - t_1) & \cdots & p(t_n - t_1) \\
0 & 1 & p(t_3 - t_2) & \cdots & p(t_n - t_2) \\
\cdots & \cdots & \cdots & \cdots & \cdots \\
0 & 0 & 0 & \cdots & p(t_n - t_{n-1})
\end{vmatrix} \geqslant 0.
$$

It is interesting to compare these with the inequality

$$
\begin{vmatrix}
1 & p(t_1) & p(t_2) & \cdots & p(t_n) \\
p(t_1) & 1 & p(t_2 - t_1) & \cdots & p(t_n - t_1) \\
\cdots & \cdots & \cdots & \cdots & \cdots \\
p(t_n) & p(t_n - t_1) & p(t_n - t_2) & \cdots & 1
\end{vmatrix} \geqslant 0
$$

which is satisfied by all p-functions since they are positive-definite.

(iii) The reader may like to prove that the family of inequalities (2.12) is essentially infinite in the sense that, for any N_0, there is a function which satisfies (2.12) for all $N \leqslant N_0$ but which is not a p-function.

(iv) Not all examples of regenerative phenomena come from Markov chains, or from the models of §2.5. Examples from queuing theory may be found in [38] and [65]. In the dam models considered for example in [29], the (indicator process of the) state of emptiness is a regenerative phenomenon, and it is interesting to notice that the cases most commonly analysed are not alternating renewal processes, but have $q = \infty$.

(v) *The sojourn time.* Let Z be a regenerative phenomenon with standard p-function p, and let Δ denote the set of dyadic rationals (numbers of the form $m2^{-n}$, where m and n are integers). Then

$$
\begin{aligned}
\mathbf{P}\{Z(t) = 1 \text{ for all } t \in (0, T) \cap \Delta\} \\
= \lim_{n \to \infty} \mathbf{P}\{Z(m2^{-n}) = 1 \ (1 \leqslant m \leqslant [2^n T])\} \\
= \lim_{n \to \infty} p(2^{-n})^{[2^n T]} \\
= e^{-qT},
\end{aligned}
$$

using (3.1). If Z is separable, this means that

$$\mathbf{P}\{Z(t) = 1 \ (0 < t < T)\} = e^{-qT}.$$

Thus, if τ denotes the sojourn time

$$\tau = \inf\{t > 0; Z(t) = 0\},$$

then

$$\mathbf{P}(\tau > T) = e^{-qT}.$$

If $q = \infty$, this implies that

$$\mathbf{P}(\tau = 0) = 1,$$

but if $q < \infty$, then τ has a negative exponential distribution with mean q^{-1}.

(vi) *The delay.* A similar analysis applies to the variable

$$T = \inf\{t > 0; Z(t) = 1\}$$

in a delayed regenerative phenomenon; T has distribution ϕ.

(vii) *Kolmogorov's theorem.* If i and j are two states in a Markov chain, then

$$p_{ij} \in \mathscr{L}(p_{jj}).$$

Corollary 1 to Theorem 2.8 therefore shows that

$$\lim_{t \to \infty} p_{ij}(t)$$

exists for all i, j. For properties of this limit see [18] or [6].

(viii) A dual concept to that of a delayed regenerative phenomenon has been discussed by Yamada [75].

(ix) *Doubly stochastic Poisson processes.* Let $\Lambda = (\Lambda(t); t \in T)$ be a non-negative stochastic process on a subset T of the real line. A doubly stochastic Poisson process with rate Λ is a point process on T which, conditional on Λ, is a non-homogeneous Poisson process for which the expected number of points in a set A is

$$\int_A \Lambda(t) \, dt.$$

The question was posed in [40]; when is a doubly stochastic Poisson process a renewal process? The answer is that this occurs if and only if

$$\Lambda(t) = \lambda Z(t),$$

where λ is a constant and Z a regenerative phenomenon. Such a renewal process has renewal density

$$h(t) = \lambda p(t).$$

Thus $\lambda p(t)$ is a renewal density for all $\lambda > 0$, if $p \in \mathscr{P}$. Daley [9] has shown that functions in $\mathscr{P}$ are the only ones with $p(0) = 1$ having this property. Since p is a renewal density, the limit theorem 2.6 is an example of a local form of Blackwell's theorem.

(x) *Renewal processes from Markov chains.* Let X be a Markov chain and a a fixed state. It has been remarked already that

$$\mathscr{S}_a = \{t; X(t) = a\}$$

is not a set of isolated points, so that ordinary renewal theory is not immediately applicable. However, the situation can be retrieved. If a is stable in the sense that

$$q_a = -p'_{aa}(0)$$

is finite, then it is known ([6], a suitable version of X being taken) that $\mathscr{S}_a$ is a union of disjoint intervals:

$$\mathscr{S}_a = (\alpha_1, \beta_1) \cup (\alpha_2, \beta_2) \cup \ldots,$$

where $\alpha_1 < \beta_1 < \alpha_2 < \beta_2 < \ldots \to \infty$. It is then possible to show that the $(\alpha_n - \alpha_{n-1})$ are independent and identically distributed, so that the entrance instants into the state a form a renewal process.

Even in this case, however, a special feature of (α_n) needs to be noted in any deep analysis, since

$$(\alpha_n - \alpha_{n-1}) = (\alpha_n - \beta_{n-1}) + (\beta_{n-1} - \alpha_{n-1}),$$

and the two summands are independent, the latter having a negative exponential distribution.

The situation is more difficult when $q_a = \infty$, for then [6] $\mathscr{S}_a$ is of positive measure but nowhere dense. A renewal process can however be found by strengthening the definition of entrance instant. Choose another state b, and say that t is an entrance instant if $X(t+) = a$ and there exists $s < t$ with $X(s) = b$ and $X(u) \neq a$ $(s < u < t)$. With this definition, the entrance instants form an isolated sequence, whose differences are independent and identically distributed.

(xi) *Croftian theory.* In proving Theorem 2.6 we used the uniform continuity of p to deduce the existence of

$$\lim_{t \to \infty} p(t)$$

from that of

$$\lim_{n \to \infty} p(nh)$$

for all $h > 0$. In deeper problems, for instance those concerned with rates of convergence, the uniformity is not known *a priori*. It is therefore useful to know the following theorem of which a particular case was discovered by Croft [8].

If f is a continuous function on $(0, \infty)$, *and if*

$$\lim_{n \to \infty} f(nh)$$

exists for all h in some set of the second category, then

$$\lim_{t \to \infty} f(t)$$

exists.

For various applications to stochastic analysis, see [36] and [33].

(xii) Both $\mathcal{Q}$ and $\mathcal{PM}$ are closed under multiplication, and are therefore sub-semigroups of $\mathcal{P}$.

(xiii) *Delphic theory.* The arithmetical structure of the semigroup $\mathcal{P}$ has been extensively studied by Kendall [33] and Davidson [11], [12], [13]. In particular, Kendall has shown that the infinitely divisible elements of $\mathcal{P}$ are exactly those of the form (5.15). He has gone on to prove that every p in $\mathcal{P}$ admits one (and in general more than one) factorisation

$$p = p_0 p_1 p_2 \cdots,$$

where $p_1, p_2, \ldots$ are indecomposable, and p_0 is infinitely divisible and has no indecomposable factors. Davidson has conjectured that p_0 must be of the form $e^{-\alpha t}$ for some $\alpha \geqslant 0$.

These properties of $\mathcal{P}$ bear a striking resemblance to those of the convolution semigroup of probability measures on the line. This led Kendall to define the notion of a delphic semigroup, which contains these as well as other interesting examples.

(xiv) *The topology of* $\mathcal{P}$. In the course of developing the delphic theory of $\mathcal{P}$, Kendall and Davidson unearthed important topological properties of $\mathcal{P}$. We have seen that $\mathcal{P}$ is not compact; the relatively compact sets in $\mathcal{P}$ turn out to be exactly those which are equicontinuous at $t = 0$. The product topology of $[0, 1]^{(0, \infty)}$ is not metrisable, but Davidson shows that the subspace topology of $\mathcal{P}$ is metrisable. He does this by showing that, on $\mathcal{P}$, it is equivalent to the compact-open topology.

(xv) The renewal sequence $u = (u_n)$ defined by

(1) $$u_n = (\lambda n)^n \, e^{-\lambda n}/n!$$

encountered in §2.5 has some interesting properties. Let $X_1, X_2, \ldots$ be independent Poisson variables with mean λ, and write

$$S_n = X_1 + X_2 + \ldots + X_n \; (S_0 = 0).$$

Then

$$Y_n = S_n - n$$

defines a Markov chain Y on the integers, and u is the renewal sequence of the state 0, since

$$\mathbf{P}(Y_n = 0) = \mathbf{P}(S_n = n) = u_n.$$

The corresponding sequence (f_n) is given by

$$f_n = \mathbf{P}\{S_r \neq r \, (r = 1, 2, \ldots, n-1), S_n = n\}$$
$$= u_n \mathbf{P}\{S_r \neq r \, (r = 1, 2, \ldots, n-1)|S_n = n\}.$$

Now, given $S_n = n$, the conditional joint distribution of $X_1, X_2, \ldots, X_n$ is independent of λ; it is in fact the symmetric multinomial distribution

$$\mathbf{P}\{X_r = x_r \, (1 \leqslant r \leqslant n)|S_n = n\} = n!/n^n x_1! x_2! \ldots x_n!.$$

Thus the problem of calculating the conditional probability f_n/u_n is a purely combinatorial one, and is solved by the classical ballot theorems [71]; it turns out to be simply $1/n$. Thus

(2) $$f_n = u_n/n \qquad (n \geqslant 1).$$

Equation (1.1.14) is for this case a consequence of the well-known identity

$$n^n = \sum_{r=1}^{n} \binom{n}{r} r^{r-1}(n-r)^{n-r}.$$

The property (2) is characteristic of the renewal sequence (1), since it implies that

$$U(z) = 1 + zF'(z),$$

so that F satisfies the differential equation

$$\frac{1 - F(z)}{F(z)} F'(z) = \frac{1}{z}.$$

This has solution

(3) $$F(z) \, e^{-F(z)} = \lambda z,$$

where λ is arbitrary, and this determines F, and so u, uniquely for a given value of λ.

(xvi) *Ergodic theory.* In the measure space $(\Omega, \mathfrak{F}, \mathbf{P})$ of Theorem 2.9, a measure-preserving transformation θ_t may be defined so that

$$Z_s(\theta_t \omega) = Z_{s+t}(\omega).$$

Thus to every p-function there corresponds a flow $(\Omega, \mathfrak{F}, \mathbf{P}, \theta_t)$; a measure space with a one-parameter group of measure-preserving transformations. Let us say that two p-functions are *E-equivalent* if their corresponding flows are isomorphic. Very little is known about this relation of equivalence of p-functions, but it appears to have some connection with the behaviour of $p(t)$ for large t. For example, Rudolfer [66] has proved a result equivalent to the assertion that, if p, p_1 and p_2 belong to $\mathscr{P}$, and if p_1 and p_2 are E-equivalent, then

$$\int_0^\infty p(t)p_1(t)\,dt \qquad \text{and} \qquad \int_0^\infty p(t)p_2(t)\,dt$$

converge or diverge together.

(xvii) *Reversibility.* Let Z be a standard regenerative phenomenon, and T a fixed positive number. Then, under the conditional probability measure

$$\mathbf{P}^T = \mathbf{P}(\cdot\,|Z(T) = 1),$$

the process $(Z(t); 0 < t < T)$ has structure determined by the equations

$$\mathbf{P}^T\{Z(t_r) = 1\ (1 \leqslant r \leqslant n)\}$$
$$= p(t_1)p(t_2 - t_1) \ldots p(t_n - t_{n-1})p(T - t_n)/p(T),$$

for $0 < t_1 < t_2 < \ldots < t_n < T$. Now define

$$\bar{Z}(t) = Z(T - t).$$

Then

$$\mathbf{P}^T\{\bar{Z}(t_r) = 1\ (1 \leqslant r \leqslant n)\} = \mathbf{P}^T\{Z(T - t_r) = 1\ (1 \leqslant r \leqslant n)\}$$
$$= p(T - t_n)p(t_n - t_{n-1}) \ldots p(t_2 - t_1)p(t_1)/p(T)$$
$$= \mathbf{P}^T\{Z(t_r) = 1\ (1 \leqslant r \leqslant n)\}.$$

In other words, Z and $\bar{Z}$ have the same joint distributions in $0 < t < T$, under the conditional measure $\mathbf{P}^T$. This reversibility property will become important in §4.3.

(xviii) For $p \in \mathscr{P}$, $\mathscr{L}(p)$ contains the constant function c if and only if $0 \leqslant c \leqslant p(\infty)$.

(xix) *Exponential decay.* For $p \in \mathscr{P}$, we have seen that the function $\phi(t) = -\log p(t)$ is non-negative, finite and subadditive. The standard theorems already quoted [24] are useful for large t as well as for small t. Thus if

$$\beta = \inf_{t>0} \phi(t)/t,$$

then

(3) $$\beta = \lim_{t\to\infty} \phi(t)/t,$$

and of course

(4) $$p(t) \leqslant e^{-\beta t}.$$

Clearly β plays the same role for p as the parameter ρ of Theorem 1.4 does for a renewal sequence. It is for instance true that $p(t)\,e^{\beta t}$ belongs to $\mathscr{P}$. (Cf. [35], [38].)

(xx) If the limit q in (3.1) is finite, then letting $s \to 0$ in (2.15) shows that

(5) $$p'(t) \geqslant -qp(t).$$

CHAPTER 3

The Theory of p-Functions

3.1 STATEMENT OF THE THEOREMS

The main questions left open by the theory developed in Chapter 2 are: what functions are p-functions, and what are their properties? These questions will now be answered for standard p-functions, and the non-standard case will be discussed in §3.4.

It was shown in §2.5 that, if $a \geqslant 0$ and B is any distribution function on $(0, \infty]$, then there is a standard p-function determined uniquely by its Laplace transform

$$\int_0^\infty p(t)\,\mathrm{e}^{-\theta t}\,dt = \left\{\theta + a - a \int_0^\infty \mathrm{e}^{-\theta x}\,dB(x)\right\}^{-1} \qquad (\theta > 0).$$

If we write μ for the measure, of total mass a, defined by

$$\mu(A) = a \int_A dB(x),$$

this equation may be written*

$$(1) \qquad \int_0^\infty p(t)\,\mathrm{e}^{-\theta t}\,dt = \left\{\theta + \int_{(0,\infty]} (1 - \mathrm{e}^{-\theta x})\mu(dx)\right\}^{-1}.$$

Thus (1) defines a function $p \in \mathscr{P}$ whenever μ is a totally finite measure on $(0, \infty]$.

One might perhaps hope that every function in $\mathscr{P}$ might admit a representation of this form, but such hopes are dashed by the observation (perhaps most easily derived from (2.5.5)) that

$$q = -p'(0) = a.$$

---

* Measures defined on the real line will always, by convention, be (positive) Borel measures. The function $\mathrm{e}^{-\theta x}$ is, when $\theta > 0$, to take the value 0 at $x = +\infty$.

Thus (1) can represent at most those p-functions with $q < \infty$, and not, for example, the function

$$p(t) = e^{-t^\eta} \qquad (0 < \eta < 1),$$

which belongs to $\mathscr{P}$ by virtue of the remarks at the end of §2.5.

It should be noted, however, that (1) sometimes makes sense even when μ has infinite total mass. Since for any $\theta > 0$ there are positive constants m_θ, M_θ such that

$$m_\theta(1 - e^{-x}) \leqslant 1 - e^{-\theta x} \leqslant M_\theta(1 - e^{-x}),$$

the integral on the right-hand side of (1) converges for all $\theta > 0$ if and only if

$$(2) \qquad \int_{(0,\infty]} (1 - e^{-x})\mu(dx) < \infty.$$

This condition implies that $\mu(\epsilon, \infty] < \infty$ for all $\epsilon > 0$, but since the integrand is small when x is small, it does not imply that $\mu(0, \epsilon] < \infty$. Thus there are measures μ of infinite total mass for which the right hand side (1) is a well-defined positive function of $\theta > 0$. It is by no means clear that this function is a Laplace transform, still less the Laplace transform of a standard p-function, but it is, and the following fundamental theorem holds.

Theorem 3.1. *If p is a standard p-function, there exists a unique measure μ on $(0, \infty]$ satisfying (2) and such that, for all $\theta > 0$,*

$$(3) \qquad \int_0^\infty p(t)\, e^{-\theta t}\, dt = \left\{ \theta + \int_{(0,\infty]} (1 - e^{-\theta x})\mu(dx) \right\}^{-1}.$$

Conversely, if μ is any measure on $(0, \infty]$ which satisfies (2), then there exists a unique continuous function p satisfying (3) for all $\theta > 0$, and p is a standard p-function.

The measure μ is called the *canonical measure* of the p-function p. Equation (3) sets up a one-to-one correspondence between $\mathscr{P}$ and the class of measures satisfying (2). Under this correspondence the standard topology of $\mathscr{P}$ induces a topology on this class of measures.

Theorem 3.2. *If p_n $(n \geqslant 1)$ and p belong to $\mathscr{P}$, and have respective canonical measures μ_n, μ, then*

$$\lim_{n \to \infty} p_n(t) = p(t)$$

for all $t > 0$ if and only if

(4)*
$$\lim_{n \to \infty} \int \phi(x)(1 - e^{-x})\mu_n(dx) = \int \phi(x)(1 - e^{-x})\mu(dx)$$

for every bounded continuous ϕ on $(0, \infty]$.

The important quantities q (2.3.1) and $p(\infty)$ (2.4.4) are simply expressed in terms of μ.

Theorem 3.3. *For a standard p-function with canonical measure μ,*

(5)
$$q = \mu(0, \infty],$$

(6)
$$p(\infty) = \left\{1 + \int x\mu(dx)\right\}^{-1}.$$

In particular, $p(\infty) > 0$ if and only if

$$\mu\{\infty\} = 0, \quad \int_{(0,\infty)} x\mu(dx) < \infty.$$

Because of (2), the function

(7)
$$m(t) = \mu(t, \infty]$$

is finite for $t > 0$, non-increasing and right-continuous. Moreover,

$$\begin{aligned}
\int_0^1 m(t)\, dt &= \int_0^1 dt \int_{(t,\infty]} \mu(dx) \\
&= \int_{(0,\infty]} \mu(dx) \int_0^{\min(x,1)} dt \\
&= \int \min(x,1)\mu(dx) \leqslant 2 \int (1 - e^{-x})\mu(dx) < \infty,
\end{aligned}$$

so that m is integrable on every finite interval.

For any function f, we shall write $\hat{f}$ for its Laplace transform

(8)
$$\hat{f}(\theta) = \int_0^\infty f(t)\, e^{-\theta t}\, dt$$

---

* Where the range of integration is omitted from an integral with respect to μ, it is to be understood to be the whole interval $(0, \infty]$ on which μ is defined. The same convention will apply to other measures with a natural interval of definition.

if it exists. Then, for $\theta > 0$,

$$\hat{m}(\theta) = \int_0^\infty e^{-\theta} \, dt \int_{(t,\infty]} \mu(dx)$$

$$= \int \frac{1 - e^{-\theta x}}{\theta} \mu(dx),$$

and (3) may be written in the form

$$\hat{p}(\theta) = \{\theta[1 + \hat{m}(\theta)]\}^{-1},$$

or

$$\theta^{-1} - \hat{p}(\theta) = \hat{p}(\theta)\hat{m}(\theta).$$

This last equation shows that p satisfies the Volterra equation

(9)
$$1 - p(t) = \int_0^t p(t - s)m(s) \, ds.$$

A formal iterative solution of (9) has the form

(10)
$$p(t) = 1 - \int_0^t \sum_{n=1}^\infty (-1)^{n-1} m_n(s) \, ds,$$

where m_n is the n-fold convolution of m with itself. We shall show that this expression is valid, and that it contains useful information about the smoothness of p.

Theorem 3.4. *A standard p-function p has finite right and left derivatives $D_+ p(t)$, $D_- p(t)$ at each $t > 0$, and*

(11)
$$D_+ p(t) - D_- p(t) = \mu\{t\},$$

so that p is differentiable at t if and only if μ has no atom at t. If μ has no atoms in the open interval $I \subseteq (0, \infty)$, then p is continuously differentiable in I, with

(12)
$$p'(t) = \sum_{n=1}^\infty (-1)^n m_n(t).$$

Theorems 3.1 and 3.3 were announced in [37] and proved in [38]. Theorem 3.2 is due to Kendall [33], and Theorem 3.4 will be found in [41]. Another result has been part of the folklore of the subject.

Theorem 3.5. *Suppose that two functions $p_1, p_2 \in \mathscr{P}$ have canonical measures μ_1 and μ_2 which coincide on the interval $(0, T)$ and satisfy*

$$\mu_1[T, \infty] = \mu_2[T, \infty].$$

Then

$$p_1(t) = p_2(t) \qquad (0 \leqslant t \leqslant T).$$

The proofs of these theorems will take up the next two sections.

3.2 PROOF OF THE FUNDAMENTAL THEOREM

Suppose first that p belongs to $\mathscr{P}$. Theorem 2.5 shows that, for each $h > 0$, there exist numbers $f_n(h)$ $(n = 1, 2, \ldots, \infty)$ with

$$f_n(h) \geqslant 0, \quad \sum_{1 \leqslant n \leqslant \infty} f_n(h) = 1,$$

such that, for $z \in (0, 1)$,

$$\sum_{n=0}^{\infty} p(nh)z^n = \left\{ 1 - \sum_{n=1}^{\infty} f_n(h)z^n \right\}^{-1}$$

$$= \left\{ \sum_{1 \leqslant n \leqslant \infty} f_n(h)(1 - z^n) \right\}^{-1}.$$

Fix $\theta > 0$, and let $z = e^{-\theta h}$. Then by the continuity of p, and the dominated convergence theorem,

$$\hat{p}(\theta) = \int_0^{\infty} p(t)\, e^{-\theta t}\, dt$$

$$= \lim_{h \to 0} \int_0^{\infty} p([t/h]h)\, e^{-\theta t}\, dt$$

$$= \lim_{h \to 0} \sum_{h=0}^{\infty} p(nh)\theta^{-1}[e^{-\theta nh} - e^{-\theta(n+1)h}]$$

$$= \lim_{h \to 0} h \sum_{n=0}^{\infty} p(nh)\, e^{-\theta nh}$$

$$= \lim_{h \to 0} h \left\{ \sum_{1 \leqslant n \leqslant \infty} f_n(h)(1 - e^{-\theta nh}) \right\}^{-1}$$

$$= \lim_{h \to 0} \left\{ \int_{[0,\infty]} \frac{1 - e^{-\theta x}}{1 - e^{-x}}\, \lambda_h(dx) \right\}^{-1},$$

where the measure λ_h on the compact interval $[0, \infty]$ has atoms

$$\lambda_h\{nh\} = h^{-1}f_n(h)(1 - e^{-nh}) \qquad (1 \leqslant n \leqslant \infty).$$

Since $\hat{p}(\theta) > 0$, this means that, for all $\theta > 0$,

(1) $$\hat{p}(\theta)^{-1} = \lim_{h \to 0} \int_{[0,\infty]} \frac{1 - e^{-\theta x}}{1 - e^{-x}}\, \lambda_h(dx).$$

In particular, taking $\theta = 1$,

$$\hat{p}(1)^{-1} = \lim_{h \to 0} \lambda_h[0, \infty],$$

showing that the total mass of λ_h is bounded as $h \to 0$.

The integrand

$$k(x, \theta) = \frac{1 - e^{-\theta x}}{1 - e^{-x}}$$

is continuous in $0 \leqslant x \leqslant \infty$ if we take

$$k(0, \theta) = \theta, \qquad k(\infty, \theta) = 1.$$

Moreover, if two totally finite measures λ and λ' satisfy

(2) $$\int k(x, \theta)\lambda(dx) = \int k(x, \theta)\lambda'(dx)$$

for all $\theta > 0$, then the fact that

$$k(x, \theta + 1) - k(x, \theta) = e^{-\theta x}$$

shows that

$$\int e^{-\theta x}\lambda(dx) = \int e^{-\theta x}\lambda'(dx),$$

so that $\lambda = \lambda'$ on $[0, \infty)$ (by classical Laplace transform theory [73]). Moreover, putting $\theta = 1$ in (2), we have $\lambda[0, \infty] = \lambda'[0, \infty]$, so that $\lambda = \lambda'$ on $[0, \infty]$.

These facts about the kernel k enable us to invoke the theorems of the Appendix to show that λ_h converges weakly, as $h \to 0$, to a measure λ on $[0, \infty]$, and that

(3) $$\hat{p}(\theta)^{-1} = \int \frac{1 - e^{-\theta x}}{1 - e^{-x}} \lambda(dx),$$

for all $\theta > 0$.

Since

$$\lim_{t \to 0} p(t) = 1,$$

we have

$$\lim_{\theta \to \infty} \theta \hat{p}(\theta) = 1,$$

so that

$$\lim_{\theta \to \infty} \int \frac{1 - e^{-\theta x}}{\theta(1 - e^{-x})} \lambda(dx) = 1,$$

and the monotone convergence theorem implies that

$$\lambda\{0\} = 1.$$

Hence

$$\hat{p}(\theta)^{-1} = \theta + \int_{(0,\infty]} \frac{1 - e^{-\theta x}}{1 - e^{-x}} \lambda(dx)$$

$$= \theta + \int_{(0,\infty]} (1 - e^{-\theta x})\mu(dx),$$

where

$$\mu(dx) = (1 - e^{-x})^{-1}\lambda(dx)$$

defines a measure μ on $(0, \infty]$ satisfying (1.2). Hence (1.3) is proved.

If μ' is another measure satisfying (1.3), then

$$\lambda'(dx) = (1 - e^{-x})\mu'(dx),$$

$$\lambda'\{0\} = 1$$

defines a measure λ' satisfying (2), so that $\lambda' = \lambda$ and $\mu' = \mu$. Hence μ is uniquely determined by p, and the first half of Theorem 3.1 is proved.

Conversely, suppose that μ is a measure satisfying (1.2), and define m by (1.7). Then

(4) $$\hat{m}(\theta) = \int_0^\infty m(t) e^{-\theta t} dt$$

exists for all $\theta > 0$, and

(5) $$\theta\hat{m}(\theta) = \int (1 - e^{-\theta x})\mu(dx).$$

Moreover,

(6) $$\lim_{\theta \to \infty} \hat{m}(\theta) = 0,$$

and we may fix $\alpha > 0$ such that

(7) $$\hat{m}(\alpha) < 1.$$

Now consider the series

(8) $$b(t) = \sum_{n=1}^\infty (-1)^{n-1}m_n(t),$$

where m_n is the n-fold convolution of m with itself. Since

$$\sum_{n=1}^\infty \int_0^\infty m_n(t) e^{-\alpha t} dt = \sum_{n=1}^\infty \hat{m}(\alpha)^n < \infty,$$

Fubini's theorem shows that (8) is absolutely convergent for almost all t, and that, for $\theta \geqslant \alpha$,

$$\hat{b}(\theta) = \sum_{n=1}^{\infty} (-1)^{n-1} \hat{m}(\theta)^n$$
$$= \hat{m}(\theta)/\{1 + \hat{m}(\theta)\}.$$

Thus the function

(9) $$p(t) = 1 - \int_0^t b(s) \, ds$$

is continuous, and has Laplace transform (in $\theta \geqslant \alpha$) equal to

$$\hat{p}(\theta) = \theta^{-1}\{1 - \hat{b}(\theta)\}$$
$$= \{\theta + \theta \hat{m}(\theta)\}^{-1}$$
$$= \left\{\theta + \int (1 - e^{-\theta x}) \mu(dx)\right\}^{-1}.$$

Thus (9) defines a continuous function p satisfying (1.3), at least for $\theta \geqslant \alpha$. Moreover, by Lerch's theorem [73], p is uniquely determined by (1.3) for $\theta \geqslant \alpha$.

If μ is totally finite, it has already been remarked in the last section that there is a function in $\mathscr{P}$ satisfying (1.3), and by the uniqueness this must be the same as the function (9). More generally, if μ is any measure satisfying (1.2), the measure μ^ϵ ($\epsilon > 0$) defined by

$$\mu^\epsilon(A) = \mu\{A_\cap(\epsilon, \infty]\}$$

is totally finite, so that there exists p^ϵ in $\mathscr{P}$ such that

(10) $$\hat{p}^\epsilon(\theta) = \left\{\theta + \int (1 - e^{-\theta x}) \mu^\epsilon(dx)\right\}^{-1}$$
$$= \left\{\theta + \int_{(\epsilon, \infty]} (1 - e^{-\theta x}) \mu(dx)\right\}^{-1}.$$

With the obvious notation,

$$m^\epsilon(t) = m[\max(\epsilon, t)],$$

so that

$$m^\epsilon(t) \leqslant m(t), \qquad \lim_{\epsilon \to 0} m^\epsilon(t) = m(t),$$

and by induction

$$m_n^\epsilon(t) \leqslant m_n(t), \qquad \lim_{\epsilon \to 0} m_n^\epsilon(t) = m_n(t).$$

Thus the dominated convergence theorem shows that

$$p(t) = 1 - \int_0^t \sum_{n=1}^{\infty} (-1)^{n-1} m_n(t)$$

$$= 1 - \lim_{\epsilon \to 0} \int_0^t \sum_{n=1}^{\infty} (-1)^{n-1} m_n^{\epsilon}(t)$$

$$= \lim_{\epsilon \to 0} p^{\epsilon}(t).$$

Hence p, a limit of p-functions, is a p-function, and (9) shows that p is standard. By dominated convergence again, (10) shows that, for $\theta > 0$,

$$\hat{p}(\theta) = \lim_{\epsilon \to 0} \hat{p}^{\epsilon}(\theta)$$

$$= \left\{ \theta + \int (1 - e^{-\theta x}) \mu(dx) \right\}^{-1},$$

and the proof of Theorem 3.1 is complete.

3.3 MORE PROOFS

Proof of Theorem 3.2. Suppose first that $p_n, p \in \mathscr{P}$ and that

$$\lim_{n \to \infty} p_n(t) = p(t)$$

for all $t > 0$. Then the dominated convergence theorem implies that

$$\lim_{n \to \infty} \hat{p}_n(\theta) = \hat{p}(\theta),$$

so that

$$\int (1 - e^{-\theta x}) \mu_n(dx) \to \int (1 - e^{-x}) \mu(dx)$$

for all $\theta > 0$. In the notation of §3.2,

$$\int k(x, \theta) \lambda_n(dx) \to \int k(x, \theta) \lambda(dx),$$

and since

$$\int k(x, \theta) \lambda(dx)$$

determines λ, this implies that λ_n converges weakly to λ, which in turn implies (1.4).

Suppose conversely that (1.4) holds. Taking

$$\phi(x) = k(x, \theta),$$

we have

(1) $$\lim_{n \to \infty} \hat{p}_n(\theta) = \hat{p}(\theta).$$

Since

$$\sum_{n=1}^{\infty} \int_0^{\infty} m_n(t)\, e^{-\alpha t}\, dt = \sum_{n=1}^{\infty} \hat{m}(\alpha)^n < \infty,$$

Fubini's theorem and (2.9) show that

(2) $$\int_0^{\infty} e^{-\alpha t}\, dp(t) \leqslant \frac{\hat{m}(\alpha)}{1 - \hat{m}(\alpha)} < \infty$$

so long as $\hat{m}(\alpha) < 1$, i.e. so long as $\alpha \hat{p}(\alpha) > \tfrac{1}{2}$. Because of (1), we can first choose α so that

$$\alpha \hat{p}(\alpha) > \tfrac{3}{4},$$

and then choose N so that

$$\alpha \hat{p}_n(\alpha) > \tfrac{2}{3} \qquad (n \geqslant N).$$

Then (2) shows that, for all $n \geqslant N$,

$$\int_0^{\infty} e^{-\alpha t}\, dp_n(t) \leqslant 2.$$

Hence ([73], Theorem 16.3) every subsequence of (p_n) has a convergent subsequence, and (1) shows that the limit of any such subsequence is p. Hence $p_n \to p$. ◆

Proof of Theorem 3.3. Theorem 3.1 having been proved, we may use the Volterra equation (1.9) to give

$$q = \lim_{t \to 0} t^{-1}\{1 - p(t)\}$$
$$= \lim_{t \to 0} t^{-1} \int_0^t p(t - s)m(s)\, ds$$
$$= m(0) = \mu(0, \infty].$$

Since

$$p(\infty) = \lim_{t \to \infty} p(t)$$

exists,

$$p(\infty) = \lim_{\theta \to 0} \theta \hat{p}(\theta)$$

$$= \lim_{\theta \to 0} \{1 + \hat{m}(\theta)\}^{-1}$$

$$= \left\{1 + \int_0^\infty m(t)\, dt\right\}^{-1}$$

$$= \left\{1 + \int x\mu(dx)\right\}^{-1},$$

with the convention that the integral is infinite if $\mu\{\infty\} > 0$. ◆
To prove Theorem 3.4 we first need an estimate for the convolution $m_n(t)$.

Lemma. *For any $\delta > 0$ and any compact interval $I \subset (0, \infty)$,*

(3) $$m_n(t) = o(\delta^n)$$

as $n \to \infty$, uniformly for $t \in I$.

Proof. If λ_{n-1} denotes $(n-1)$-dimensional Lebesgue measure on the simplex

$$U_n = \{y = (y_1, y_2, \ldots, y_n) \in R^n; y_j \geq 0, \Sigma y_j = 1\},$$

then the convolution $m_n(t)$ may be written

(4) $$m_n(t) = t^{n-1} \int_{U_n} m(ty_1)m(ty_2) \ldots m(ty_n)\lambda_{n-1}(dy).$$

If

$$U_{nk} = \{y \in U_n; y_k = \max y_j\},$$

then

$$U_n = \bigcup_{k=1}^n U_{nk}$$

and

$$\lambda_{n-1}(U_{nk} \cap U_{nl}) = 0 \qquad (k \neq l),$$

so that

$$m_n(t) = t^{n-1} \sum_{k=1}^n \int_{U_{nk}} m(ty_1)m(ty_2) \ldots m(ty_n)\lambda_{n-1}(dy)$$

$$= nt^{n-1} \int_{U_{nn}} m(ty_1)m(ty_2) \ldots m(ty_n)\lambda_{n-1}(dy),$$

by symmetry. On U_{nn}, $y_n \geqslant n^{-1}$, and since m is non-increasing,

$$m_n(t) \leqslant nt^{n-1}m(tn^{-1}) \int_{U_{nn}} m(ty_1) \ldots m(ty_{n-1})\lambda_{n-1}(dy)$$

$$\leqslant nt^{n-1}m(tn^{-1}) \int_{U_n} m(ty_1) \ldots m(ty_{n-1})\lambda_{n-1}(dy).$$

In this integral, make the change of variable

$$(y_1, y_2, \ldots, y_n) \to (s, z_1, z_2, \ldots, z_{n-1}),$$

where

$$s = t(y_1 + y_2 + \ldots + y_{n-1}), \qquad z_j = ty_j/s.$$

Then

$$m_n(t) \leqslant nt^{n-1}m(tn^{-1}) \int_0^t (s/t)^{n-1} \, ds \int_{U_{n-1}} m(sz_1) \ldots m(sz_{n-1})\lambda_{n-2}(dz)$$

$$= nm(tn^{-1}) \int_0^t ds \, m_{n-1}(s)$$

$$\leqslant nm(tn^{-1}) \int_0^t e^{\theta(t-s)}m_{n-1}(s) \, ds$$

$$\leqslant nm(tn^{-1}) \, e^{\theta t}\hat{m}(\theta)^{n-1},$$

for any $\theta > 0$. Thus, if $I = [a, b]$ ($0 < a < b < \infty$), then

(5) $$\sup_I m_n(t) \leqslant nm(an^{-1}) \, e^{\theta b}\hat{m}(\theta)^{n-1}.$$

Choose θ so that $\hat{m}(\theta) < \delta$ and note that, since m is non-increasing,

$$xm(x) \leqslant \int_0^x m(t) \, dt \to 0$$

as $x \to 0$. Hence the right hand side of (5), as $n \to \infty$, is

$$o\{n^2\hat{m}(\theta)^{n-1}\} = o(\delta^n). \qquad \blacklozenge$$

Proof of Theorem 3.4. The lemma shows that the series (2.8) converges absolutely, uniformly in every compact interval of $(0, \infty)$. Moreover, if $n \geqslant 2$, the integrand in (4) is continuous almost everywhere modulo λ_{n-1}, since m has at most a countable number of discontinuities. Hence m_n is continuous for $n \geqslant 2$ (monotone convergence) and therefore

$$b_1(t) = \sum_{n=2}^{\infty} (-1)^{n-1}m_n(t)$$

is continuous, and

$$b(t) = m(t) + b_1(t).$$

Now m is discontinuous only at the atoms of μ, so that b has only jump discontinuities, and

$$b(t+) = m(t+) + b_1(t)$$
$$b(t-) = m(t-) + b_1(t)$$

for all $t > 0$. Hence (2.9) shows that p has finite right and left derivatives

$$D_{\pm} p(t) = -b(t\pm),$$

and

$$D_+ p(t) - D_- p(t) = m(t-) - m(t+) = \mu\{t\}.$$

If, in particular, μ has no atoms in an interval I, m and hence b are continuous in I, and (2.9) shows that p has continuous derivative b in I. ◆

Proof of Theorem 3.5. Under the conditions of the theorem, $m_1 = m_2$ on $(0, T)$ and (2.9) then shows that $p_1 = p_2$ on $[0, T]$.   ◆

3.4 NON-STANDARD p-FUNCTIONS

Theorem 3.1, by associating each standard p-function with a measure satisfying the convergence condition (1.2), describes the class $\mathscr{P}$ fairly explicitly. The p-functions which fail to satisfy the condition

$$\lim_{t \to 0} p(t) = 1$$

are much more difficult to handle, and their properties remained obscure for some years. Eventually the following theorem was proved, and this section will be devoted to sketching its proof, omitting some technical points for which the reader is referred to [48].

Theorem 3.6. *Let p be a Lebesgue measurable p-function. Then either*
  (i) *there is a number a in $0 < a \leqslant 1$ and a standard p-function $\bar{p}$ with*

(1) $$p(t) = a\bar{p}(t),$$

or (ii) $p(t) = 0$ *for almost all t.*

Both possibilities can occur. Indeed, let $Z(t)$ be independent variables with

$$\mathbf{P}\{Z(t) = 1\} = a, \qquad \mathbf{P}\{Z(t) = 0\} = 1 - a.$$

Then (2.2.1) is satisfied with

(2) $$p(t) = a \qquad (t > 0).$$

Thus (2) is a p-function, and by Theorem 2.2 (1) is also a p-function for all $\bar{p} \in \mathscr{P}$. We have already encountered (in §2.6) examples of p-functions which vanish except on the integers, and more complex examples may be given (§3.6(xiii)).

Sketch of proof. Suppose that the p-function p is Lebesgue measurable, and that $\{t; p(t) > 0\}$ has positive measure, so that

$$\hat{p}(\theta) = \int_0^\infty p(t)\,e^{-\theta t}\,dt$$

exists and is positive for all $\theta > 0$. We have to prove that p is of the form (1).

We first note that, if g is any probability density on $(0, \infty)$, and F is defined by (2.2.11), then the numbers

$$f_n = \int \cdots \int_{0 < t_1 < t_2 < \ldots < t_n} F(t_1, t_2, \ldots, t_n; p)$$

$$\times g(t_1)g(t_2 - t_1) \ldots g(t_n - t_{n-1})\,dt_1 \ldots dt_n$$

satisfy

$$f_n \geq 0, \qquad \sum_{n=1}^\infty f_n \leq 1.$$

The renewal sequence associated with (f_n) is computed to be

$$(3) \qquad u_0 = 1, \qquad u_n = \int_0^t p(t)g_n(t)\,dt,$$

where g_n is the n-fold convolution of g with itself. (This is a sort of generalisation of Theorem 2.5.) In particular, taking $g(t) = \alpha\,e^{-\alpha t}$ we see that, for any $\alpha > 0$,

$$(4) \qquad u_0 = 1, \qquad u_n = \int_0^\infty \frac{\alpha^n t^{n-1}}{(n-1)!}\,e^{-\alpha t}p(t)\,dt$$

defines a renewal sequence.

Equation (2.3.9) shows that

$$p_\alpha(t) = \sum_{n=0}^\infty u_n \pi_n(\alpha t)$$

defines a *standard* p-function p_α. This has Laplace transform

$$\hat{p}_\alpha(\theta) = \sum_{n=0}^{\infty} u_n \alpha^n (\alpha + \theta)^{-n-1}$$

$$= \frac{1}{\alpha + \theta} + \int_0^\infty \sum_{n=1}^\infty \frac{\alpha^n t^{n-1}}{(n-1)!} e^{-\alpha t} p(t) \frac{\alpha^n}{(\alpha + \theta)^{n+1}} dt$$

$$= \frac{1}{\alpha + \theta} + \frac{\alpha^2}{(\alpha + \theta)^2} \int_0^\infty \exp\left[-\alpha t + \frac{\alpha^2 t}{\alpha + \theta} \right] p(t)\, dt$$

$$= \frac{1}{\alpha + \theta} + \left(\frac{\alpha}{\alpha + \theta} \right)^2 \hat{p}\left(\frac{\alpha \theta}{\alpha + \theta} \right),$$

so that

$$\hat{p}(\theta) = \lim_{\alpha \to \infty} \hat{p}_\alpha(\theta).$$

Since p_α belongs to $\mathscr{P}$, there exists (2.3) a totally finite measure λ_α on $[0, \infty]$ such that

$$\hat{p}_\alpha(\theta)^{-1} = \int k(x, \theta) \lambda_\alpha(dx).$$

Letting $\alpha \to \infty$, and using the same weak convergence arguments as in §3.2, we deduce the existence of a probability measure λ on $[0, \infty]$ such that

(5) $$\hat{p}(\theta)^{-1} = \int k(x, \theta) \lambda(dx),$$

where the positivity of $\hat{p}(1)$ is used to infer that the total mass of λ_α is bounded as $\alpha \to \infty$.

Since $p(t) \leqslant 1$, $\hat{p}(\theta) \leqslant \theta^{-1}$, so that

$$\int \frac{1 - e^{-\theta x}}{\theta(1 - e^{-x})} \lambda(dx) \geqslant 1;$$

letting $\theta \to \infty$ we have

$$\lambda\{0\} \geqslant 1.$$

We may therefore define

$$a = \lambda\{0\}^{-1}, \qquad 0 < a \leqslant 1,$$

$$\mu(A) = a \int_A (1 - e^{-x})^{-1} \lambda(dx), \qquad A \subseteq (0, \infty].$$

Then (5) becomes

$$\hat{p}(\theta) = a\left\{\theta + \int_{(0,\infty]} (1 - e^{-\theta x})\mu(dx)\right\}^{-1}$$

$$= a\int_0^\infty \bar{p}(t)\, e^{-\theta t}\, dt$$

by Theorem 3.1, where $\bar{p} \in \mathscr{P}$ is the standard p-function with canonical measure μ. Since this holds for all $\theta > 0$, we have

(6) $$p(t) = a\bar{p}(t) \text{ for almost all } t.$$

Our task is now to remove the qualification 'almost', bearing in mind that we do not yet know that p is continuous.

Define independent regenerative phenomena Z and Z^a with respective p-functions $\bar{p}$ and a. Then

$$\tilde{Z}(t) = Z(t)Z^a(t)$$

defines a regenerative phenomenon $\tilde{Z}$ with p-function

$$\tilde{p}(t) = a\bar{p}(t);$$

$\tilde{p}$ is continuous, and $p = \tilde{p}$ almost everywhere.

Consider now the inequality

$$F(t_1, t_2, \ldots, t_n; p) \geqslant 0.$$

Keep $t_n = T$ fixed, and let $t = (t_1, t_2, \ldots, t_{n-1})$ move over the open set

$$G_T = \{t \in R^{n-1}; 0 < t_1 < t_2 < \ldots < t_{n-1} < T\}.$$

From the form of F,

$$F(t, T; p) - p(T) = F(t, T; \tilde{p}) - \tilde{p}(T)$$

for almost all $t \in G_T$, modulo Lebesgue measure in R^{n-1}. Thus

$$F(t, T; \tilde{p}) - \tilde{p}(T) + p(T) \geqslant 0$$

for almost all $t \in G_T$. But this last expression is continuous in t, so that the inequality holds for all $t \in G_T$, and so

$$\tilde{p}(T) - p(T) \leqslant F(t, T; \tilde{p})$$
$$= \mathbf{P}\{\tilde{Z}(t_1) = \tilde{Z}(t_2) = \ldots = \tilde{Z}(t_{n-1}) = 0, \tilde{Z}(t_n) = 1\}.$$

An exactly similar analysis, using

$$\sum_{r=1}^n F(t_1, \ldots, t_r; p) \leqslant 1,$$

shows that

$$p(T) - \tilde{p}(T) \leqslant \mathbf{P}\{\tilde{Z}(t_1) = \tilde{Z}(t_2) = \ldots = \tilde{Z}(t_n) = 0\},$$

so that

$$
\begin{aligned}
|p(T) - \tilde{p}(T)| &\leqslant \mathbf{P}\{\tilde{Z}(t_1) = \tilde{Z}(t_2) = \ldots = \tilde{Z}(t_{n-1}) = 0\} \\
&= \mathbf{P}\{Z^a(t_r) = 0 \text{ whenever } Z(t_r) = 1\} \\
&= \mathbf{E}\{(1 - a)^N\},
\end{aligned}
$$

where N is the number of values of r in $1 \leqslant r \leqslant n - 1$ for which $Z(t_r) = 1$.
Now Z is standard, and it is easy to see that, if N_k is the value of N corresponding to

$$n = 2^k, \qquad t_r = r2^{-k}T,$$

then N_k increases to infinity as $k \to \infty$. The monotone convergence theorem shows that

$$\mathbf{E}\{(1 - a)^{N_k}\} \to 0,$$

so that

$$p(T) = \tilde{p}(T) = a\bar{p}(T).$$

Since T is arbitrary, the theorem is proved.   ◆

3.5   INEQUALITIES

In his work [21] on Markov chains, Freedman has found a need for an answer to the following general question. If $0 < s < t$, and if the value of $p_{ii}(t)$ is known for some state i, how small can $p_{ii}(s)$ be? With Blackwell he has produced a partial answer in the form of the following inequality ([4], independently discovered by Davidson [12]):
If $p_{ii}(t) > \frac{3}{4}$, then for $0 < s < t$,

$$p_{ii}(s) \geqslant \tfrac{1}{2}\{1 + [4p_{ii}(t) - 3]^{\frac{1}{2}}\}.$$

In the language of §2.1, this asserts an inequality satisfied by all $p \in \mathscr{PM}$. But $\mathscr{PM}$ is dense in $\mathscr{P}$, so that the inequality must hold for all $p \in \mathscr{P}$; it is worth giving a direct proof.

Theorem 3.7. *If the standard p-function p satisfies $p(t) > \frac{3}{4}$ for some t, then for $0 < s < t$,*

(1) $$p(s) \geqslant \tfrac{1}{2}\{1 + [4p(t) - 3]^{\frac{1}{2}}\}.$$

Proof. Write $x = p(s)$, $y = p(t - s)$, $z = p(t)$, and allow s to vary with t fixed. Then the point (x, y) describes a continuous curve in the unit square, from the point $(1, z)$ to the point $(z, 1)$. The right hand inequality of (2.2.15) restricts (x, y) to lie in the set

$$K_z = \{(x, y); 0 \leqslant x, y \leqslant 1, x(1 - y) \leqslant 1 - z, (1 - x)y \leqslant 1 - z\}.$$

It is easily verified (draw a figure) that, when $z > \frac{3}{4}$, K_z has two connected components, one contained in the square $0 \leqslant x, y \leqslant \eta$ and the other in the square $\xi \leqslant x, y \leqslant 1$, where $\eta < \xi$ and η, ξ are the roots of the quadratic

$$\xi(1 - \xi) = 1 - z.$$

Hence (x, y) cannot escape from the component containing $(1, z)$, and since $z \geqslant \xi$ this is the component in $\xi \leqslant x, y \leqslant 1$. Hence

$$x \geqslant \xi = \tfrac{1}{2}\{1 + [4z - 3]^{\frac{1}{2}}\},$$

which is equivalent to (1). ◆

This inequality is far from being sharp, and indeed Davidson, in a posthumous note, showed how it could be sharpened, for some values of $p(t)$, by using a remarkable inequality due to Bloomfield [5].

Theorem 3.8. *If p is a standard p-function with canonical measure μ, then*

$$(2) \qquad p(t) \geqslant \exp\left\{-\int \min(t, x)\mu(dx)\right\}.$$

Proof. Suppose first that $p(\infty) > 0$, and apply (1.6.7) to the renewal sequence $(p(nh))$ to give

$$p(nh) \geqslant \exp\{\log p(h)[p(\infty)^{-1} - 1]/[1 - p(h)]\}.$$

Set $h = t/n$ and let $n \to \infty$, keeping t fixed, to give

$$(3) \qquad p(t) \geqslant \exp[1 - p(\infty)^{-1}].$$

More generally, if p is any standard p-function with canonical measure μ, let $\bar{p}$ be the p-function with canonical measure defined by

$$\bar{\mu}(A) = \mu(A) \qquad (A \subseteq (0, t))$$
$$\bar{\mu}(A) = 0 \qquad (A \subseteq (t, \infty])$$
$$\bar{\mu}\{t\} = \mu[t, \infty].$$

Then Theorem 3.5 implies that

$$p(t) = \bar{p}(t)$$
$$\geqslant \exp\left[1 - \bar{p}(\infty)^{-1}\right]$$
$$= \exp\left[-\int x\bar{\mu}(dx)\right]$$
$$= \exp\left\{-\int_{(0,t]} x\mu(dx) - t\mu(t, \infty]\right\}$$
$$= \exp\left\{-\int \min(t, x)\mu(dx)\right\}. \qquad \blacklozenge$$

*Corollary.** Let $p \in \mathscr{P}$, $t > 0$ and

(4) $$a = \inf\{p(s); 0 \leqslant s \leqslant t\}.$$

Then

(5) $$p(t) \leqslant \tfrac{3}{4} \qquad (a < \tfrac{1}{2})$$
$$p(t) \leqslant 1 + a \log a \qquad (0 < a \leqslant 1).$$

Proof. If $p(t) > \tfrac{3}{4}$, (1) shows that $p(s) \geqslant \tfrac{1}{2}$ and $a \geqslant \tfrac{1}{2}$. It is therefore only necessary to prove the second inequality of (5). Since p is continuous, we may choose a value of $s \in (0, t]$ such that $p(s) = a$. Write

$$l(t) = \int_0^t m(s)\, ds = \int_{(0,\infty]} \min(t, x)\mu(dx),$$

so that Bloomfield's inequality (2) is

$$p(t) \geqslant e^{-l(t)}.$$

By (1.9),

$$1 - p(t) = \int_0^t p(t - s)m(s)\, ds$$
$$\geqslant a \int_0^t m(s)\, ds$$
$$= al(t)$$
$$\geqslant al(s)$$
$$\geqslant a \log(1/a),$$

which proves (5). $\blacklozenge$

---

*This result was reconstructed by David Williams from the notes left by Rollo Davidson.

The second inequality is stronger than the first if and only if $a > \eta$, where $\eta = 0 \cdot 116$ satisfies $\eta \log \eta = -\frac{1}{4}$. There is no reason to suppose that (5) is best possible, and it would be interesting to find the sharp upper bound

$$(6) \qquad \vartheta(a) = \sup \{p(t); p \in \mathscr{P}, p \text{ satisfies } (4)\}.$$

The example (2.5.6) shows (after a little algebra) that

$$(7) \qquad \vartheta(a) \geqslant e^{a-1},$$

and it is not beyond the bounds of possibility that there is equality in (7). In any event, (5) and (7) determine $\vartheta(a)$ approximately when a is near 1:

$$(8) \qquad \vartheta(a) = a + \tfrac{1}{2}(1 - a)^2 + O\{(1 - a)^3\}.$$

The more interesting case is that in which a is small; if p becomes very small on $(0, t)$, how large can $p(t)$ be? It seems likely that $\vartheta(a)$ converges to a limit ϑ_0 as $a \to 0$, and if so we must have

$$(9) \qquad e^{-1} \leqslant \vartheta_0 \leqslant \tfrac{3}{4}.$$

The problem of finding ϑ is one of the more important unsolved problems of the theory of p-functions, but it might more naturally be formulated in a rather more general way, that of describing the set

$$(10) \qquad \Theta(t/s) = \{(p(s), p(t)); p \in \mathscr{P}\}.$$

This is a subset of the unit square, and depends only on the ratio t/s since a p-function remains a p-function if the time scale is changed. Then inequalities like (1) have the effect of delimiting Θ.

Although for no value of λ except $\lambda = 1$ is $\Theta(\lambda)$ known exactly, a number of facts are known, and are set down here in the hope that they may later fall into place in a satisfactory treatment. We always take $\lambda > 1$, without loss of generality.

(a) If (x_1, y_1), (x_2, y_2) belong to $\Theta(\lambda)$, so does $(x_1 x_2, y_1 y_2)$;
(b) $\Theta(\lambda)$ contains all points (x, y) with

$$(11) \qquad x^\lambda \leqslant y \leqslant x;$$

(c) if λ is an integer, every point $(x, y) \in \Theta(\lambda)$ satisfies

$$(12) \qquad x^\lambda \leqslant y.$$

The reason for (b) is that any such point can be realised with a p-function of the form (2.5.15). If it is taken with (c) it describes the part of $\Theta(\lambda)$

lying in $y \leqslant x$, so long as λ is an integer (but not otherwise [70]). The part of $\Theta(\lambda)$ lying in $y > x$ is more difficult to study, since it arises from p-functions which are not monotonic.

The set $\Theta(\lambda)$ is not closed, but its closure is

(13) $$\overline{\Theta}(\lambda) = \{(p(s), p(t)); p \in \overline{\mathscr{P}}\},$$

where $\overline{\mathscr{P}}$ is the closure of $\mathscr{P}$ in $[0, 1]^{(0, \infty)}$, which is of course compact. The elements of $\overline{\mathscr{P}}$ are all p-functions, but not every p-function is in $\overline{\mathscr{P}}$. For example, the inequality (1) holds for all $p \in \overline{\mathscr{P}}$, so that for instance the p-function

$$p(t) = 1 \quad (t \text{ integral})$$
$$= 0 \quad (\text{otherwise})$$

does not belong to $\overline{\mathscr{P}}$. In fact, any p-function in $\overline{\mathscr{P}}$ which is almost everywhere zero satisfies

(14) $$p(t) \leqslant \vartheta_0 \leqslant \tfrac{3}{4}$$

for all $t > 0$. The problems of studying $\vartheta(a)$ for small a, and examining $\Theta(\lambda)$ near $x = 0$, are very closely connected with that of deciding which p-functions, almost everywhere zero, belong to $\overline{\mathscr{P}}$. None of them is easy (cf. [49], [52]).

3.6 NOTES

(i) The measure μ of Theorem 3.1 can be expressed directly in terms of the numbers $f_n(h)$. If $0 < a < b < \infty$ and if neither a nor b are atoms of μ, then

(1) $$\mu(a, b) = \lim_{h \to 0} h^{-1} \sum_{a < nh < b} f_n(h).$$

(ii) It follows easily from (1.3) that, if the variable θ is regarded as complex, then

(2) $$\operatorname{Re} \hat{p}(\theta) > 0 \quad (\operatorname{Re} \theta > 0).$$

Moreover [38], $\hat{p}$ has a continuous extension to

$$\{\theta; \operatorname{Re} \theta \geqslant 0, \theta \neq 0\},$$

and the density g in (2.6.11) may be written

$$(3) \qquad\qquad g(\omega) = 2\pi^{-1} \operatorname{Re}\{\hat{p}(i\omega)\}.$$

(iii) If $p \in \mathscr{Q}$ is of the form (2.3.9), its canonical measure may easily be expressed in terms of the sequence (f_n) associated with the renewal sequence (u_n). It has a density

$$(4) \qquad\qquad h(t) = \sum_{n=2}^{\infty} f_n \lambda^2 \pi_{n-2}(\lambda t)$$

and an atom $\mu\{\infty\} = \lambda f_\infty$.

(iv) The full converse of Theorem 2.5 is false, since the function

$$
\begin{aligned}
p(t) &= e^{-t} && (t \text{ rational}) \\
&= e^{-2t} && (t \text{ irrational})
\end{aligned}
$$

has $(p(nh)) \in \mathscr{R}$ for all $h > 0$. It cannot be a p-function, since if it were it would be standard, and could not fail to be continuous. However, the argument of §3.2 can easily be made to show that, *if a function p continuous on $[0, \infty)$ has $(p(nh); n \geqslant 0) \in \mathscr{R}$ for arbitrarily small values of h, then p belongs to $\mathscr{P}$.*

(v) *Completely monotonic functions.* For any probability measure v on $[0, \infty)$, the completely monotonic function [73]

$$(5) \qquad\qquad p(t) = \int e^{-xt} v(dx)$$

has $-\log p(t)$ continuous and concave, and so belongs to $\mathscr{P}$. It is shown in [44] (the result in one direction being due to Reuter [64]) that these p-functions are exactly those whose canonical measures have completely monotonic densities on $(0, \infty)$. Moreover, they all belong to $\mathscr{P}\mathscr{M}$, and indeed are exactly the diagonal transition functions which can arise from *reversible* Markov chains.

(vi) *The rate of convergence problem.* The results of this chapter, and particularly (1.3), have been used [38] to throw light on the rate of convergence of $p(t)$ to $p(\infty)$, a topic of some importance in Markov theory. A useful result is that the abscissa of regularity [73] of the Laplace transform of $p(t) - p(\infty)$ coincides with its abscissa of convergence.

(vii) *Bounded variation.* It was shown in §3.3 that, for large α, the function $p(t) e^{-\alpha t}$ has bounded variation on $(0, \infty)$. A much stronger result is proved in [41], namely that $p(t)$ itself has bounded variation;

$$(6) \qquad\qquad \int_0^\infty |p'(t)| \, dt < \infty,$$

so long as $p(\infty) > 0$. Indeed, it is quite possible that this is true even when $p(\infty) = 0$, since no counterexample is known.

A related question is whether, for all $p \in \mathscr{P}$,

(7)
$$\lim_{t \to 0} p'(t) = 0.$$

This is known [6] to be true when $p \in \mathscr{PM}$, and Sykes [70] has used the identity

(8)
$$p'(t) = -m(t)p(t) - \int_{[0,t]} [p(t) - p(t - x)]\mu(dx)$$

to show that (7) holds also if

$$\int_0^1 m(t)^2 \, dt < \infty.$$

An argument will be described in §6.6 which, as a by-product, establishes (7) whenever $p(\infty) > 0$.

(viii) *Behaviour at the origin.* It is important to note that the continuous differentiability of p, when μ is non-atomic, has only been proved for $t > 0$. When q is finite, it is easy to show that

$$\lim_{t \to 0} p'(t) = -q,$$

but this is not always true when q is infinite. Indeed, Smith [67] has produced a function in $\mathscr{PM}$ with

$$\lim_{t \to 0} \inf p'(t) = -\infty,$$

$$\lim_{t \to \infty} \sup p'(t) = +\infty.$$

However, Kendall has shown that, near the origin, $p'(t)$ is usually negative in the sense that, if $W_+(t)$ and $W_-(t)$ are the positive and negative variations of p on $[0, t]$, then

(9)
$$W_+(t) \leqslant \frac{2W_-(t)^2}{1 - 2W_-(t)}$$

when $W_-(t) < \frac{1}{2}$. His proof [30] is apparently restricted to $p \in \mathscr{PM}$, but it applies to the general case $p \in \mathscr{P}$ without essential change. The weaker fact that

$$W_+(t) = O[W_-(t)^2]$$

as $t \to 0$ may easily be deduced from (1.12).

Another simple property of $p(t)$ for small t is a direct consequence of (1.9):

(10)
$$1 - p(t) \sim \int_0^t m(s) \, ds$$

as $t \to 0$. This incidentally shows that Bloomfield's inequality is asymptotically sharp as $t \to 0$.

(ix) *The interpolation problem.* Davidson has proposed the problem of characterising those renewal sequences u for which $u_n = p(n)$ for some p in $\mathscr{P}$ (or $\overline{\mathscr{P}}$). This problem, which seems difficult, has some affinities with the 'imbedding problem' for Markov chains [34].

(x) *A binary operation on measures.* Let μ_1 and μ_2 be measures satisfying (1.2), and p_1, p_2 the corresponding p-functions. Then $p = p_1 p_2$ is a standard p-function, and its canonical measure μ may be computed, in principle, in terms of μ_1 and μ_2. Write

$$\mu = \mu_1 \# \mu_2,$$

so that $\#$ is an associative binary operation on the class M of measures satisfying (1.2). The properties of the semigroup $(M, \#)$ are of course the same as that of the isomorphic semigroup $\mathscr{P}$. However, the operation $\#$ is extremely obscure, and any light thrown upon it would be welcome.

(xi) It is a suggestive fact that the condition (2.5.14) for the function

$$\tilde{p}(t) = \exp \left\{ -\int \min (t, x) \mu(dx) \right\}$$

to belong to $\mathscr{P}$ is exactly equivalent to the condition (1.2) for μ to be the canonical measure of a p-function p, since

$$1 - e^{-x} \leqslant \min (1, x) \leqslant 2(1 - e^{-x}).$$

One connection between p and $\tilde{p}$ has already been noted; Bloomfield's inequality $p(t) \geqslant \tilde{p}(t)$. Another is given by a pair of convergence results, easily proved by Theorem 3.2.

(a) Let p_n be the p-function with canonical measure $n^{-1}\mu$. Then

$$\lim_{n \to \infty} p_n(t)^n = \tilde{p}(t).$$

(b) Let μ_n be the canonical measure of $\tilde{p}^{1/n}$. Then $n\mu_n$ converges to μ in the sense of (1.4).

(**xii**) The function p_α used in §3.4 has the explicit expression

$$(11) \qquad p_\alpha(t) = e^{-\alpha t} + \alpha \int_0^\infty s^{-\frac{1}{2}} t^{\frac{1}{2}} e^{-\alpha(s+t)} I_1\{2\alpha(st)^{\frac{1}{2}}\} p(s)\, ds,$$

where I_1 is the usual Bessel function.

(**xiii**) *Examples of p-functions which have zero values.* Let G be any proper subgroup of the additive group of real numbers, and define Z_G as the indicator function of $G \cap (0, \infty)$. Then (2.2.1) is trivially satisfied, with p-function

$$\begin{aligned} p_G(t) &= 1 \qquad (t > 0, t \in G) \\ &= 0 \qquad \text{(otherwise)}. \end{aligned}$$

These are in fact the only p-functions which take both values 0 and 1, and no others.

For any $a \in (0, 1]$ and $\tilde{p} \in \mathscr{P}$,

$$(12) \qquad\qquad\qquad p(t) = a p_G(t) \tilde{p}(t)$$

defines a p-function with

$$(13) \qquad\qquad \{t > 0; p(t) > 0\} = G \cap (0, \infty).$$

It would be interesting to know whether there are groups G such that every (measurable?) p-function satisfying (13) is of the form (12). This is certainly not true, for example, for the group of integers.

Another class of examples arises as follows. Let C be any countable set of positive reals, and f a non-negative function on C with

$$\sum_C f(c) \leqslant 1.$$

Then it is possible to verify that the expression

$$(14) \qquad\qquad p(t) = \sum_{n=1}^\infty \sum_{c_1+c_2+\ldots+c_n\,=t} f(c_1)f(c_2)\ldots f(c_n)$$

is a p-function, which vanishes off the additive semigroup generated by C. This construction admits further generalisation [48].

If G is non-measurable, then p_G is non-measurable, so that not all p-functions are measurable. It has been conjectured that all strictly positive p-functions are Lebesgue measurable (and hence by Theorem 3.6 continuous in $(0, \infty)$).

(**xiv**) *An inversion formula.* Let the standard p-function p have canonical measure μ. Let ϕ be a function on $(0, \infty)$ which is integrable on finite intervals, and write

(15) $$\Phi(t) = \int_0^t p(t - s)\phi(s)\, ds.$$

This convolution transform can be inverted by the formula

(16) $$\int_0^t \phi(s)\, ds = \Phi(t) + \int_0^t \Phi(t - u)m(u)\, du,$$

which is an immediate consequence of the Volterra equation (3.1.9) [45].

(**xv**) Corollary 2 to Theorem 2.8 is characteristic of $\mathscr{P}$; if a continuous function $p: [0, \infty) \to [0, 1]$ satisfies (2.6.6) for all $a, t > 0$, when $\phi(\cdot, a)$ is a probability measure for each a, then p belongs to $\mathscr{P}$.

CHAPTER 4

Sample Function Properties of Regenerative Phenomena

4.1 VERSIONS OF A REGENERATIVE PHENOMENON

In this section, p will denote a fixed standard p-function. By definition, a regenerative phenomenon with p-function p is a stochastic process $(Z(t); t > 0)$ satisfying the equation

(1)
$$\mathbf{P}\{Z(t_r) = 1 \ (1 \leqslant r \leqslant n)\} = \prod_{r=1}^{n} p(t_r - t_{r-1}),$$

whenever $0 = t_0 < t_1 < t_2 < \ldots < t_n$. More explicitly, Z is a family of measurable functions $Z(t): \Omega \to \{0, 1\}$ on a probability space $(\Omega, \mathfrak{F}, \mathbf{P})$, satisfying (1). To be pedantic, therefore, a phenomenon is a quartet

$$\Omega, \quad \mathfrak{F}, \quad \mathbf{P}, \quad Z,$$

and there are many phenomena for a given p.

It is convenient to regard Z as a function from Ω into the product space

(2)
$$\Pi = \{0, 1\}^{(0, \infty)}$$

whose elements are the functions f from $(0, \infty)$ into the two-point set $\{0, 1\}$. Thus, for $\omega \in \Omega$, $Z\omega$ is the function in Π which sends $t > 0$ into

$$Z(t)(\omega) = Z(t, \omega).$$

In the usual way, Z induces a probability measure $\mathbf{P}_Z$ on the σ-algebra

(3)
$$\mathfrak{B}_Z = \{A \subseteq \Pi; Z^{-1}A \in \mathfrak{F}\}$$

---

* The reader whose interest is restricted to the properties of p-functions and the characterisation of $\mathscr{P}\mathscr{M}$ will find that the remaining chapters do not depend significantly on the arguments of this chapter.

in Π, by the equation

(4) $$\mathbf{P}_Z(A) = \mathbf{P}(Z^{-1}A).$$

To every regenerative phenomenon Z is therefore associated the probability space $(\Pi, \mathfrak{B}_Z, \mathbf{P}_Z)$. Let Φ_t denote the subset of Π given by

$$\Phi_t = \{f \in \Pi; f(t) = 1\}.$$

Then the defining condition (1) simply asserts that

$$Z^{-1} \bigcap_{r=1}^{n} \Phi_{t_r}$$

belongs to $\mathfrak{F}$, and has probability given by the right hand side of (1). In other words, $\mathfrak{B}_Z$ must contain every set of the form

$$A = \bigcap_{r=1}^{n} \Phi_{t_r},$$

and $\mathbf{P}_Z$ must satisfy

$$\mathbf{P}_Z(A) = \prod_{r=1}^{n} p(t_r - t_{r-1}).$$

If $\mathfrak{B}_0$ is the smallest σ-algebra of subsets of Π containing the sets Φ_t $(t > 0)$, then the argument of §2.2 shows that $\mathfrak{B}_Z$ contains $\mathfrak{B}_0$, and that the restriction of $\mathbf{P}_Z$ to $\mathfrak{B}_0$ is a probability measure $\mathbf{P}_0$ uniquely determined by p.

To express this in less abstract terms, the following question can be answered from a knowledge of the function p alone, provided that A belongs to $\mathfrak{B}_0$.

What is the probability that a sample function of a regenerative phenomenon with p-function p lies in the subset A of Π?

The answer is just $\mathbf{P}_0(A)$, and depends only on the p-function, and not on any further properties of the phenomenon.

The area of uniqueness may be extended in a number of directions. In the first place, it is usual to assume that, in the probability space $(\Omega, \mathfrak{F}, \mathbf{P})$, $\mathfrak{F}$ is complete for $\mathbf{P}$ (i.e. that $E \subseteq F$, $F \in \mathfrak{F}$, $\mathbf{P}(F) = 0$ implies $E \in \mathfrak{F}$). If this is so, then the restriction of $\mathbf{P}_Z$ to the completion $\mathfrak{B}$ of $\mathfrak{B}_0$ under $\mathbf{P}_0$ is uniquely determined by $\mathbf{P}_0$, and hence by p. Thus, whatever phenomenon Z we use, we must have

(5) $$\mathbf{P}(Z \in A) = \mathbf{P}_Z(A) = \mathbf{P}_p(A),$$

so long as $A \in \mathfrak{B}$, where $\mathbf{P}_p$ is determined uniquely by p. (Notice that $\mathfrak{B}$, unlike $\mathfrak{B}_0$, depends on p via $\mathbf{P}_0$.)

To go further, regularity conditions must be imposed on the process Z. For instance, if Z is assumed to be separable [15], there will be events such as

$$\bigcap_{a<t<b} \Phi_t$$

which may not be in $\mathfrak{B}$ but whose probabilities are nonetheless determined uniquely by p. The events with this property form a σ-algebra $\mathfrak{B}_S$. Similarly, the events whose probabilities under $\mathbf{P}_Z$ are the same for all measurable regenerative phenomena Z form a σ-algebra $\mathfrak{B}_M$. Those whose probabilities are the same for all separable measurable phenomena with given p-function form a σ-algebra $\mathfrak{B}_{SM}$. Those whose probabilities are the same for all such right-continuous phenomena form a σ-algebra $\mathfrak{B}_{RC}$, and so on. We have the inclusions

(6)
$$\mathfrak{B}_0 \subset \mathfrak{B} \quad \begin{array}{c} \subset \mathfrak{B}_S \subset \\ \\ \subset \mathfrak{B}_M \subset \end{array} \quad \mathfrak{B}_{SM} \subset \mathfrak{B}_{RC},$$

which are strict except in trivial cases.

The importance of these rather arcane remarks is as follows. Suppose we construct, as we shall in the next section, a regenerative phenomenon Z with given p-function which has the property that its sample functions are right-continuous. Suppose we then prove that a certain property of the sample functions of Z is true with probability P. That is, we prove that

$$\mathbf{P}(Z \in A) = P$$

for some subset A of the set Π of possible sample functions. Then, if $A \in \mathfrak{B}$, it will be true that

(7)
$$\mathbf{P}(Z \in A) = P$$

for *all* regenerative phenomena with the given p-function. Similarly, if $A \in \mathfrak{B}_S$, (7) is true for all such separable phenomena. If $A \in \mathfrak{B}_{SM}$, (7) is true for all separable measurable phenomena with the given p-function, and so on.

Similar remarks hold of course for much more general stochastic processes, but they are particularly apposite here because a very convenient right-continuous version has been constructed by D. G. Kendall (in lectures at Cambridge which one may hope will not long remain unpublished, cf. [3]). His construction is described in detail in the next section, but was suggested by the following property.

If Z is a measurable regenerative phenomenon with the given standard p-function, then

$$(8) \qquad U(t) = \int_0^t Z(u) \, du$$

is a well-defined stochastic process, whose sample functions are continuous and non-decreasing. Thus U has a right-continuous inverse function

$$(9) \qquad V(t) = \inf \{s > 0; \ U(s) > t\},$$

where as usual $\inf \phi = +\infty$. A tedious argument was used in [38] to show that V is a process with stationary, non-negative, independent increments, whose distributions are determined by the formula

$$(10) \qquad \mathbf{E}\{e^{-\theta V(t)}\} = \exp\{-t\hat{p}(\theta)^{-1}\}, \qquad (\theta > 0).$$

Kendall's idea is to reverse this argument, starting with V, to construct Z.

4.2 KENDALL'S CONSTRUCTION

If μ is any measure on $(0, \infty]$ satisfying the usual finiteness condition

$$(1) \qquad \int (1 - e^{-x})\mu(dx) < \infty,$$

then it is well known [58] that there exists a right-continuous, strong Markov process $(V_0(t); t \geq 0)$ $(V_0(0) = 0)$ with non-negative, stationary, independent increments, and satisfying

$$(2) \qquad \mathbf{E}\{e^{-\theta V_0(t)}\} = \exp\left\{-t \int (1 - e^{-\theta x})\mu(dx)\right\}.$$

If $\mu\{\infty\} > 0$, there is a finite ζ such that

$$V_0(t) < \infty \ (t < \zeta), \qquad V_0(\zeta) = \infty;$$

if $\mu\{\infty\} = 0$ we set $\zeta = \infty$. The process V_0 increases only in jumps, and for any $\epsilon > 0$ the positions of the jumps of height $> \epsilon$ form a Poisson process of rate $\mu(\epsilon, \infty)$ on $[0, \zeta]$, their heights being independent with distribution function

$$(3) \qquad \mu(\epsilon, x)/\mu(\epsilon, \infty).$$

If V_0 is such a process, then

$$(4) \qquad V(t) = t + V_0(t)$$

defines another additive process with similar properties, except that it increases at unit rate between its jumps. The process Z is then defined by

(5) $Z(t) = 1$ (if $V(s) = t$ for some $s > 0$)

               $= 0$ (otherwise).

Clearly this defines a right-continuous stochastic process with state space $\{0, 1\}$.

Theorem 4.1. The process Z defined by (5) is a standard regenerative phenomenon whose p-function has canonical measure μ.

Proof. For $t > 0$, write $E(t)$ for the event

$$E(t) = \{Z(t) = 1\} = \{V(s) = t \text{ for some } s > 0\}.$$

Fix $0 = t_0 < t_1 < t_2 < \ldots < t_n$, and confine attention to the subset $E(t_1)$ of the probability space Ω. If this has positive probability, we can consider the conditional probability measure $\mathbf{P}_1 = \mathbf{P}\{\cdot \,|E(t_1)\}$. Since V is strictly increasing, the value of s with $V(s) = t_1$ is unique, and is a stopping time for the strong Markov process V. Hence, under $\mathbf{P}_1$, the process

$$V_1(t) = V(s + t) - t_1 \ (t \geqslant 0)$$

is a copy of V itself, so that

$$\mathbf{P}_1\{ \bigcap_{r=2}^{n} E(t_r)\} = \mathbf{P}_1\{V(s_r) = t_r \text{ for some } s_r \ (2 \leqslant r \leqslant n)\}$$

$$= \mathbf{P}_1\{V_1(u_r) = t_r - t_1 \text{ for some } u_r \ (2 \leqslant r \leqslant n)\}$$

$$= \mathbf{P}\{V(u_r) = t_r - t_1 \text{ for some } u_r \ (2 \leqslant r \leqslant n)\}$$

$$= \mathbf{P}\{ \bigcap_{r=2}^{n} E(t_r - t_1)\}.$$

Writing

$$p(t) = \mathbf{P}\{E(t)\},$$

we therefore have

$$\mathbf{P}\{ \bigcap_{r=1}^{n} E(t_r)\} = p(t_1)\mathbf{P}\{ \bigcap_{r=2}^{n} E(t_r - t_1)\}$$

if $p(t_1) > 0$, and the equation is also trivially true of $p(t_1) = 0$.

A trivial induction gives

$$\mathbf{P}\{ \bigcap_{r=1}^{n} E(t_r)\} = \prod_{r=1}^{n} p(t_r - t_{r-1}),$$

which shows that Z is a regenerative phenomenon with p-function p, since

$$E(t) = \{Z(t) = 1\}.$$

It remains to identify the p-function. Since Z is right-continuous, it is a measurable process, and Fubini's theorem applies to give, for $\theta > 0$,

$$
\begin{aligned}
\hat{p}(\theta) &= \int_0^\infty e^{-\theta t} \mathbf{E}\{Z(t)\}\, dt \\
&= \mathbf{E} \int_0^\infty e^{-\theta t} Z(t)\, dt \\
&= \mathbf{E} \int_{t;V(s)=t \text{ for some } s} e^{-\theta t}\, dt \\
&= \mathbf{E} \int_0^\infty e^{-\theta V(s)}\, ds \\
&= \int_0^\infty \mathbf{E}\{e^{-\theta s - \theta V_0(s)}\}\, ds \\
&= \int_0^\infty \exp\left\{-\theta s - s \int (1 - e^{-\theta x})\mu(dx)\right\} ds \\
&= \left\{\theta + \int (1 - e^{-\theta x})\mu(dx)\right\}^{-1}.
\end{aligned}
$$

Hence p must be the standard p-function with canonical measure μ. ◆

The structure of the resulting phenomenon is reasonably easy to understand. The parameter set $(0, \infty)$ is divided into the random subsets

$$\mathscr{T}_1 = \{t > 0; Z(t) = 1\}, \qquad \mathscr{T}_0 = \{t > 0; Z(t) = 0\}.$$

Then $\mathscr{T}_0$ corresponds to the jumps of V; it is a countable disjoint union of half-open intervals. The same is true of $\mathscr{T}_1$ only when q is finite; when $q = \infty$ its structure is more complex.

The arguments described in §4.1 enable general sample function properties of regenerative phenomena to be read off from the corresponding results for the Kendall version. A typical theorem is the following result, to which allusion has already been made, and which was proved more directly in [38].

Theorem 4.2. *Let Z be any measurable regenerative phenomenon with standard p-function p, and define*

(6) $$U(t) = \int_0^t Z(u)\, du,$$

(7) $$V(t) = \inf\{s > 0; U(s) > t\}.$$

Then V is a process with stationary independent increments, and

(8)
$$\mathbf{E}\{e^{-\theta V(t)}\} = \exp\{-t\hat{p}(\theta)^{-1}\}.$$

Proof. In the notation of §4.1, $U(t)$, regarded as a function of $Z \in \Pi$, is $\mathfrak{B}_M$-measurable by Fubini's theorem. Since

$$\{V(t) < v\} = \bigcup_{\substack{0 < s < v \\ s \text{ rational}}} \{U(s) > t\},$$

$V(t)$ is $\mathfrak{B}_M$-measurable. Therefore, as in §4.1, the finite-dimensional distributions of V are the same for every measurable phenomenon with the given p-function, and it suffices to prove the result for the version constructed in Theorem 4.1.

To avoid confusion, denote the additive process (4) from which that construction starts by $\bar{V}$. Then

$$U(t) = \int_0^t Z(u)\,du$$
$$= |\{s; \bar{V}(s) < t\}|$$

if $|\cdot|$ denotes Lebesgue measure. Thus

$$V(t) = \bar{V}(t),$$

using the right-continuity of $\bar{V}$, and so V has stationary independent increments and

$$\mathbf{E}\{e^{-\theta V(t)}\} = \mathbf{E}\{e^{-\theta \bar{V}(t)}\}$$
$$= e^{-\theta t}\mathbf{E}\{e^{-\theta V_0(t)}\}$$
$$= \exp\left\{-t\left[\theta + \int (1 - e^{-\theta x})\mu(dx)]\right]\right\}$$
$$= \exp\{-t\hat{p}(\theta)^{-1}\}. \qquad \blacklozenge$$

4.3 THE FORWARD AND BACKWARD PROCESSES; JOINT DISTRIBUTION

If Z is a standard regenerative phenomenon, and t any positive number, then we may ask the question: 'at time t, how long has it been since last $Z = 1$, and how long will it be before $Z = 1$ again?' In other words, what can be said about the random variables

(1)
$$B(t) = \inf\{s \geqslant 0; Z(t - s) = 1\},$$

(2)
$$F(t) = \inf\{s \geqslant 0; Z(t + s) = 1\}?$$

It is clear that both $B(t)$ and $F(t)$ are $\mathcal{B}_s$-measurable functions of Z, so that the finite-dimensional distributions of the processes $(B(t); t > 0)$ and $(F(t); t > 0)$ are determined by the p-function of Z provided Z is separable. For the Kendall version,

(3) $$B(t) = \inf \{t - V(s); V(s) \leqslant t\},$$

(4) $$F(t) = \inf \{V(s) - t; V(s) \geqslant t\},$$

and

(5) $$\{F(t) = 0\} = \{Z(t) = 1\},$$

(6) $$\{B(t) = 0\} = \{Z(t-) = 1\}.$$

It will be shown that B and F are both Markov processes, with transition functions* which can be expressed in terms of the p-function and its canonical measure. It turns out, however, to be more natural to consider first the bivariate process

(7) $$K(t) = (B(t), F(t)),$$

which will be shown to be also a Markov process.

If the canonical measure μ has a density h in $(0, \infty)$, an important role is played by the measure in the positive quadrant of the plane whose density is $h(x + y)$. This can be defined even when μ does not have a density by the formula

(8) $$\Gamma_0(E) = \iint\limits_{(x-y,y) \in E} \mu(dx)\, dy,$$

where E runs over the Borel subsets of the quadrant $(0, \infty) \times (0, \infty)$. For some purposes, it is useful to define a measure Γ on the closed quadrant $[0, \infty) \times [0, \infty)$ by

(9) $$\Gamma = \epsilon_{(0,0)} + \Gamma_0,$$

where ϵ_a denotes the probability measure concentrated at the point a.

Lemma. *If the standard p-function p has canonical measure μ, then for s, $t > 0$,*

(10) $$p(s + t) - p(s)p(t) = \int_0^s \int_0^t p(s - u)p(t - v)\Gamma_0(du\, dv).$$

---

* For the notation used here (following Doob [15]) for Markov processes with uncountable state spaces, see Chapter 7. Notice that we may have $F(t) = \infty$ for transient phenomena.

Proof. For α, $\beta > 0$, $\alpha \neq \beta$,

$$
\begin{aligned}
\int_0^\infty \int_0^\infty e^{-\alpha u - \beta v} \Gamma_0(du\, dv) &= \iint_{0 \leqslant y \leqslant x} e^{-\alpha(x-y)-\beta y} \mu(dx)\, dy \\
&= \int_{(0,\infty)} \frac{e^{-\alpha x} - e^{-\beta x}}{\beta - \alpha}\, \mu(dx) \\
&= (\beta - \alpha)^{-1}\{\hat{p}(\beta)^{-1} - \beta - \hat{p}(\alpha)^{-1} + \alpha\}.
\end{aligned}
$$

Hence

$$
\begin{aligned}
\hat{p}(\alpha)\hat{p}(\beta) \iint e^{-\alpha u - \beta v} \Gamma_0(du\, dv) &= \frac{\hat{p}(\alpha) - \hat{p}(\beta)}{\beta - \alpha} - \hat{p}(\alpha)\hat{p}(\beta) \\
&= \int_0^\infty \int_0^\infty p(s+t)\, e^{-\alpha s - \beta t}\, ds\, dt - \hat{p}(\alpha)\hat{p}(\beta),
\end{aligned}
$$

which is the double Laplace transform of (10).   ◆

Theorem 4.3. *Let Z be a separable standard regenerative phenomenon. Then K is a (possibly dishonest) Markov process on $[0, \infty) \times [0, \infty)$, with transition function given by*

(11)
$$
\begin{aligned}
P_t((b,f); A) &= \epsilon_{(b+t, t-f)}(A) \qquad (t < f) \\
&= P_{t-f}(A) \qquad\qquad (t \geqslant f),
\end{aligned}
$$

where

(12)
$$
P_t(A) = \iint_A p(t-b)\Gamma(db\, df).
$$

Proof. There is no loss of generality in using Kendall's version of Z. Fix $T > 0$, and consider the process $\tilde{K}(t) = K(T + t)$, given that $K(T) = (b, f)$ and given information about the past $K(u)$ $(u < T)$. There is a unique τ with

$$
V(\tau-) \leqslant T \leqslant V(\tau),
$$

and τ is a stopping time for the strong Markov process V. Hence

$$
\tilde{V}(t) = V(\tau + t) - V(\tau) \qquad (t \geqslant 0)
$$

is a process with the same distributions as V, independent of the pre-τ σ-algebra of events. Since $K(T) = (b, f)$,

$$
V(\tau-) = T - b, \qquad V(\tau) = T + f.
$$

Hence, for $t > 0$,

$$F(T + t) = \inf \{V(s) - (T + t); V(s) \geqslant T + t\}$$
$$= \inf \{\tilde{V}(u) + f - t; \tilde{V}(u) \geqslant t - f\},$$

and similarly

$$B(T + t) = \inf \{t - f - \tilde{V}(u); \tilde{V}(u) \leqslant t - f\}$$

for $t \geqslant f$, and

$$B(T + t) = b + t$$

for $t < f$. Thus, for $t < f$,

$$K(T + t) = (b + t, f - t),$$

while for $t \geqslant f$,

$$K(T + t) = \tilde{K}(t - f),$$

where $\tilde{K}$ is defined in terms of $\tilde{V}$ in the same way that K is defined in terms of V. This shows that K is a Markov process, and that its transition function is of the form (11), where

$$P_t(A) = P_t((0, 0); A).$$

Note that

$$P_t\{(0, 0)\} = p(t).$$

By the Chapman–Kolmogorov equation,

$$P_{s+t}(A) = P_{s+t}((0, 0); A)$$

$$= \iint P_s((0, 0); db\,df)P_t((b, f); A)$$

$$= \iint_{f \leqslant t} P_s(db\,df)P_{t-f}(A) + \iint_{f > t} P_s(db\,df)\, \epsilon_{(b+t,f-t)}(A).$$

Taking Laplace transforms with respect to s and t we have, for $\lambda, \nu > 0$, $\lambda \neq \nu$,

$$\frac{\hat{P}_\lambda(A) - \hat{P}_\nu(A)}{\nu - \lambda}$$

$$= \iint \hat{P}_\lambda(db\,df)\, e^{-\nu f}\hat{P}_\nu(A) + \iint \hat{P}_\lambda(db\,df) \int_{(b+t,f-t)\in A} e^{-\nu t}\,dt.$$

Thus, if

$$\Pi(\lambda, \alpha, \beta) = \iint \hat{P}_\lambda(db\,df)\, e^{-\alpha b - \beta f},$$

we have, when $\alpha - \beta + \nu \neq 0$,

$$\{\Pi(\lambda, \alpha, \beta) - \Pi(\nu, \alpha, \beta)\}/(\nu - \lambda)$$

$$= \iint \hat{P}_\lambda(db\,df)\, e^{-\nu f}\Pi(\nu, \alpha, \beta)$$

$$+ \iint \hat{P}_\lambda(db\,df) \int_0^f e^{-\alpha(b+t)-\beta(f-t)-\nu t}\, dt$$

$$(13) \qquad = \Pi(\lambda, 0, \nu)\Pi(\nu, \alpha, \beta)$$

$$+ \{\Pi(\lambda, \alpha, \beta) - \Pi(\lambda, \alpha, \alpha + \nu)\}/(\alpha - \beta + \nu).$$

Let $\beta \to \infty$ and use the fact that

$$\lim_{\beta \to \infty} \Pi(\lambda, \alpha, \beta) = \hat{P}_\lambda\{(0, 0)\} = \hat{p}(\lambda),$$

to give

$$\frac{\hat{p}(\lambda) - \hat{p}(\nu)}{\nu - \lambda} = \Pi(\lambda, 0, \nu)\,\hat{p}(\nu).$$

Use this to substitute for $\Pi(\lambda, 0, \nu)$ in (13), to yield after a little rearrangement the equation

$$(14) \qquad \frac{\alpha - \beta + \lambda}{\hat{p}(\lambda)}\, \Pi(\lambda, \alpha, \beta)$$

$$= \frac{\alpha - \beta + \nu}{\hat{p}(\nu)}\, \Pi(\nu, \alpha, \beta) + \frac{\lambda - \nu}{\hat{p}(\lambda)}\, \Pi(\lambda, \alpha, \alpha + \nu),$$

which in this form clearly holds even if $\lambda - \nu = 0$ or $\alpha - \beta + \nu = 0$. Thus if

$$\psi(\lambda, \alpha, \beta) = \frac{\alpha - \beta + \lambda}{\hat{p}(\lambda)}\, \Pi(\lambda, \alpha, \beta),$$

then for $\lambda, \nu > 0$, $\alpha, \beta \geqslant 0$,

$$(15) \qquad \psi(\lambda, \alpha, \beta) = \psi(\nu, \alpha, \beta) + \psi(\lambda, \alpha, \alpha + \nu).$$

Writing

$$f(\alpha, \beta) = \psi(1, \alpha, \beta), \qquad g(\lambda, \alpha) = \psi(\lambda, \alpha, \alpha + 1),$$

(15) with $\nu = 1$ gives

$$\psi(\lambda, \alpha, \beta) = f(\alpha, \beta) + g(\lambda, \alpha).$$

Substituting back into (15), we then get

$$g(\nu, \alpha) = -f(\alpha, \alpha + \nu),$$

whence

$$\psi(\lambda, \alpha, \beta) = f(\alpha, \beta) - f(\alpha, \alpha + \lambda)$$

and

(16) $$\Pi(\lambda, \alpha, \beta) = \hat{p}(\lambda)\frac{f(\alpha, \beta) - f(\alpha, \alpha + \lambda)}{\alpha - \beta + \lambda}.$$

The simplest way of determining f is to use the conditional reversibility of Z established in §2.7 (xvii). This clearly implies that, given $Z(s + t) = 1$, the distribution of $(B(s), F(s))$ is the same as that of $(F(t), B(t))$. Now

$$\mathbf{E}\{e^{-\alpha B(s) - \beta F(s)} | Z(s + t) = 1\}$$

$$= p(s + t)^{-1}\int\int e^{-\alpha b - \beta f}P_s(db\ df)P_t((b, f); \{(0, 0)\})$$

$$= p(s + t)^{-1}\int\int e^{-\alpha b - \beta f}P_s(db\ df)p(t - f),$$

so that

$$\int\int e^{-\alpha b - \beta f}P_s(db\ df)p(t - f) = \int\int e^{-\beta b - \alpha f}P_t(db\ df)p(s - f).$$

Taking Laplace transforms with respect to s and t, we have

$$\Pi(\lambda, \alpha, \beta + \nu)\hat{p}(\nu) = \Pi(\nu, \beta, \alpha + \lambda)\hat{p}(\lambda),$$

which by (16) means that

(17) $$f(\alpha, \beta + \nu) - f(\alpha, \alpha + \lambda) = f(\beta, \beta + \nu) - f(\beta, \alpha + \lambda).$$

Writing $$f(\nu) = f(0, \nu),$$

$$h(\alpha) = f(\alpha, \alpha + 1) - f(0, \alpha + 1),$$

(17) with $\beta = 0$, $\lambda = 1$ gives

$$f(\alpha, \nu) = f(\nu) + h(\alpha),$$

so that

(18) $$\Pi(\lambda, \alpha, \beta) = \hat{p}(\lambda)\frac{f(\beta) - f(\alpha + \lambda)}{\alpha - \beta + \lambda}.$$

Setting $\alpha = 0$ and using the expression above for $\Pi(\lambda, 0, \nu)$,

$$f(\beta) - f(\lambda) = \frac{\hat{p}(\beta) - \hat{p}(\lambda)}{\hat{p}(\beta)\hat{p}(\lambda)},$$

so that

$$f(\beta) = -\hat{p}(\beta)^{-1}$$

(ignoring an irrelevant additive constant) and

$$\Pi(\lambda, \alpha, \beta) = \hat{p}(\lambda) \frac{\hat{p}(\alpha + \lambda)^{-1} - \hat{p}(\beta)^{-1}}{\alpha - \beta + \lambda}$$

$$= \hat{p}(\lambda) \left\{ 1 + \int \frac{e^{-\beta x} - e^{-(\alpha + \lambda)x}}{\alpha - \beta + \lambda} \mu(dx) \right\}$$

$$= \hat{p}(\lambda) \int\int e^{-(\alpha + \lambda)b - \beta f} \Gamma(db\ df).$$

Therefore

$$\hat{P}_\lambda(A) = \hat{p}(\lambda) \int\int_A e^{-\lambda b} \Gamma(db\ df),$$

and (12) follows. ◆

4.4 THE FORWARD AND BACKWARD PROCESSES; MARGINAL DISTRIBUTIONS

Theorem 4.3 determines the finite-dimensional distributions of the bivariate process $K = (B, F)$, and hence indirectly those of the univariate processes B and F. Of course, the Markovian nature of K by no means implies that B and F are Markov processes, but it turns out that the particular form of the transition function of K, expressed in (3.11) and (3.12), does imply this.

If Γ is the measure defined by (3.9), then for $B \subseteq R^+ = [0, \infty)$,

$$(1) \qquad \Gamma(B \times R^+) = \Gamma(R^+ \times B) = \gamma(B),$$

where the marginal measure γ is given by

$$(2) \qquad \gamma(B) = \epsilon_0(B) + \int\int_{x-y\in B} \mu(dx)\ dy = \epsilon_0(B) + \int_B m(u)\ du.$$

Theorem 4.4. *Under the conditions of Theorem 4.3, the process F is Markovian, with transition function P^F given by*

$$(3) \qquad P_t^F(x, A) = \epsilon_{x-t}(A) \qquad\qquad (t < x)$$

$$= \int_A p(t - x - y)\gamma(dy) \qquad (t \geqslant x).$$

(In this and similar expressions, the convention is that p vanishes for negative functions of its argument.)

Proof. Equation (3.12) may be written

$$(4) \qquad P_t(du\ dv) = p(t - u)\Gamma(du\ dv),$$

so that

$$P_t(du \times R^+) = p(t - u)\gamma(du).$$

Thus

$$\mathbf{P}\{F(s + t) \in A | K(s) = (b, f)\} = \epsilon_{f-t}(A) \qquad (t < f).$$
$$= \int_A p(t - f - y)\gamma(dy) \qquad (t \geqslant f).$$

Since this does not depend on b, and since K is a Markov process, F is Markovian, with transition function given by (3). $\blacklozenge$

In order to deal with the backward process, which is more difficult, note that the measures G_t and H_t, defined for all $t > 0$ on R^+ by

$$(5) \qquad G_t(A) = \Gamma(A \times [t, \infty))$$

and

$$(6) \qquad H_t(A) = \int_0^t p(t - \tau)\Gamma(A \times d\tau),$$

are absolutely continuous with respect to γ; we denote by

$$(7) \qquad g(b, t) = G_t(db)/\gamma(db), \qquad h(b, t) = H_t(db)/\gamma(db)$$

the corresponding derivatives, which may be supposed to have been chosen to be jointly measurable in (b, t).

Theorem 4.5. *Under the conditions of Theorem 4.3, the process B is Markovian, with transition function*

$$(8) \qquad P_t^B(x, A) = g(x, t)\epsilon_{x+t}(A) + \int_A h(x, t - u)\gamma(du).$$

Proof. By Corollary 2 to Theorem 2.8, there exists for all $a > 0$ a probability measure $\phi(\cdot, a)$ on $[0, \infty]$ such that, for all $t > 0$,

$$(9) \qquad p(t + a) = \int_0^t p(t - u)\phi(du, a).$$

By setting $\phi(\cdot, a) = \epsilon_{-a}$ for $a \leqslant 0$, (9) becomes true for all real a. By (4),

$$P_{s+t}(A) = \int\!\!\int_A p(s + t - u)\Gamma(du\,dv)$$

$$= \int\!\!\int_A \int_0^s p(s - \xi)\phi(d\xi, t - u)\Gamma(du\,dv)$$

$$= \int_0^s p(s - \xi) \int\!\!\int_A \phi(d\xi, t - u)\Gamma(du\,dv).$$

On the other hand, the Chapman–Kolmogorov equation implies that

$$P_{s+t}(A) = \int_{R^+} \int_{R^+} P_s(db\,df)P_t((b,f); A)$$

$$= \int_{R^+} p(s - b) \int_{R^+} \Gamma(db\,df)P_t((b,f); A).$$

Holding t, A fixed, comparing these two expressions for $P_{s+t}(A)$ and using the uniqueness implied by the inversion formula of §3.6 (xiv), we see that, for all $B \subseteq R^+$,

(10) $$\int\!\!\int_{B\,R^+} \Gamma(db\,df)P_t((b,f); A) = \int\!\!\int_A \phi(B, t - u)\Gamma(du\,dv).$$

This formula allows us to prove, by induction on n, that for $B_1, B_2, \ldots, B_n, C \subseteq R^+$,

$$\mathbf{P}\{B(t_r) \in B_r \,(1 \leqslant r \leqslant n), F(t_n) \in C\}$$

$$= \int_{B_1} \cdots \int_{B_nC} p(t - b_1)\prod_{r=1}^{n-1} \phi(db_{r-1}, t_r - t_{r-1} - b_r)\Gamma(db_n\,df_n).$$

In particular, with $C = R^+$,

(11) $\mathbf{P}\{B(t_r) \in B_r(1 \leqslant r \leqslant n)\}$

$$= \int_{B_1} \cdots \int_{B_2} p(t - b_1)\prod_{r=2}^{n-1} \phi(db_{r-1}, t_r - t_{r-1} - b_r)\gamma(db_n).$$

Now

$$\mathbf{P}\{B(s) \in B, B(s + t) \in C\}$$

$$= \int_B \int_{R^+} P_s(db\, df) P_t((b, f); C \times R^+)$$

$$= \int_B p(s - b) \Big\{ \int_{[t, \infty)} \Gamma(db\, df) \epsilon_{(b + t, f - t)}(C \times R^+)$$

$$+ \int_{[0, t)} \Gamma(db\, df) P_{t - f}(C \times R^+) \Big\}$$

$$= \int_B p(s - b) \Big\{ \Gamma(db \times [t, \infty)) \epsilon_{b + t}(C)$$

$$+ \int_{[0, t)} \Gamma(db\, df) \int_C p(t - f - u) \Gamma(du \times R^+) \Big\}$$

$$= \int_B p(s - b) \Big\{ G_t(db) \epsilon_{b + t}(C) + \int_C H_{t - v}(db) \gamma(du) \Big\}$$

$$= \int_B p(s - b) \gamma(db) \Big\{ g(b, t) \epsilon_{b + t}(C) + \int_C h(b, t - u) \gamma(du) \Big\}.$$

Since

$$\mathbf{P}\{B(s) \in B\} = \int_B p(s - b) \gamma(db),$$

this shows that

$$\mathbf{P}\{B(s + t) \in C \mid B(s)\} = P_t^B(B(s), C),$$

where P^B is given by (8). Notice that P_t^B does not depend on s. From (11),

$$\int_B p(s - b) \gamma(db) P_t^B(b, C) = \mathbf{P}\{B(s) \in B, B(s + t) \in C\}$$

$$= \int_B \int_C p(s - b) \phi(db, t - u) \gamma(du),$$

so that

(12) $$\gamma(db) P_t^B(b, C) = \int_C \phi(db, t - u) \gamma(du).$$

Since $\gamma(\{0\}) = 1$, this means that

$$P_t^B(0, C) = \int_C \phi(\{0\}, t - u) \gamma(du),$$

and from (9),

$$\phi(\{0\}, a) = p(a),$$

so that

(13) $$P_t^B(0, C) = \int_C p(t - u)\gamma(du).$$

Thus (12) and (13) can be used to throw (11) into the form

$$\mathbf{P}\{B(t_r) \in B_r \,(1 \leqslant r \leqslant n)\} = \int_{B_1} \cdots \int_{B_n} \prod_{r=1}^{n} P_{t_r - t_{r-1}}^B(b_{r-1}, db_r),$$

with the convention that $b_0 = t_0 = 0$. ◆

4.5 NOTES

(i) One of the advantages of Kendall's construction is that it permits the use of arguments based on the strong Markov property of V. A similar construction, raising more explicitly the possibility of a 'strong regenerative property', has been given by Hoffmann–Jørgensen [25].

(ii) *Topological versions.* The discussion of §4.1 was based on the classical Daniell–Kolmogorov approach. In [50] it was shown that, for some purposes, there are advantages in using the more topological Kakutani–Nelson technique. Thus the basic product space Π is compact in its product topology, and the standard theory of measures [22] may be invoked to show that

(a) $\mathfrak{B}_0$ is the σ-algebra of Baire sets, the smallest σ-algebra containing the compact G_δ sets, and
(b) $\mathfrak{B}_0$ is a proper subalgebra of the σ-algebra $\mathfrak{B}$. of Borel sets, the smallest σ-algebra containing the open sets, and there is exactly one regular measure on $\mathfrak{B}$. which coincides with $\mathbf{P}_0$ on $\mathfrak{B}_0$.

With this measure the phenomenon is automatically separable.

(iii) In the Markov chain case, Theorem 4.2 was known to Lévy [57], who used his theory of additive processes to gain a deep insight into the behaviour of the sample functions of Markov chains.

(iv) *Recurrence.* For a standard regenerative phenomenon, the following statements are equivalent, and define the property of recurrence.

(a) $Z(t) = 1$ for arbitrarily large values of t, with probability 1.

(b) $\int_0^\infty Z(t)\,dt = \infty$, with probability 1.

(c) For some (and then for all) $h > 0$, $Z(nh) = 1$ for infinitely many n, with probability one.

(d) $\int_0^\infty p(t)\,dt = \infty$.

(e) For some (and then for all) $h > 0$, $\sum p(nh) = \infty$.

(f) $\mu\{\infty\} = 0$.

(v) *Transience.* A standard regenerative phenomenon which is not recurrent is called transient. For a transient phenomenon, the lifetime ζ of V is finite, and indeed

$$\mathbf{P}\{\zeta > z\} = e^{-\mu\{\infty\}z}$$

for all $z > 0$. Moreover,

$$\zeta = \int_0^\infty Z(t)\,dt.$$

(vi) *Non-standard phenomena.* All the regenerative phenomena considered in this chapter have been standard, and it may be asked whether anything useful can be said about regenerative phenomena whose p-functions are not standard. Clearly if p vanishes outside a null set N, then regenerative phenomena can be constructed with p-functions p which also vanish outside N. On the other hand, not every interesting version will have this property. For example, if $W(t)$ denotes a Wiener process, and $Z(t)$ is defined by

(1) $\qquad\qquad Z(t) = 1 \qquad$ if $W(t) = 0$,

$\qquad\qquad\qquad\quad = 0 \qquad$ otherwise,

then Z is a regenerative phenomenon with p-function identically zero. Clearly this last statement ignores all the structure of the process Z (but see (xv)).

If Z is a regenerative phenomenon with p-function of the form

$$p(t) = a\bar{p}(t) \qquad (0 < a < 1, \bar{p} \in \mathscr{P}),$$

then its sample function behaviour is extremely irregular. It is impossible, for example, for Z to be measurable [48].

(vii) The known properties of V may easily be translated into theorems about the structure of the sample functions of Z. Thus for example the results summarised in proposition 14 of [38] follow without great difficulty.

(viii) *Ladder phenomena.* An interesting construction of a regenerative phenomenon has been given by Rubinovitch [65]. Let $(X(t); t \geqslant 0)$ be a process with stationary independent increments (positive or negative), and define

$$(2) \qquad Z(t) = 1 \qquad \text{if } X(s) \leqslant X(t) \text{ for all } s < t,$$
$$= 0 \qquad \text{otherwise.}$$

He shows that Z is a regenerative phenomenon, with p-function

$$(3) \qquad p(t) = \mathbf{P}\{X(u) \geqslant 0 \text{ for all } u \leqslant t\}$$

(if $X(0) = 0$). Either p is standard, or p is identically zero. In the former case, a number of interesting applications follow.

(ix) *Invariant measures.* If in (4.10) we set $B = R^+$, we obtain the equation

$$(4) \qquad \iint \Gamma(db\,df)P_t((b,f);A) = \Gamma(A),$$

which shows that Γ is an invariant measure for the Markov process K. If $p(\infty) > 0$ rather more can be said, since (3.12) then shows (using the dominated convergence theorem) that

$$(5) \qquad \lim_{t \to \infty} P_t(A) = p(\infty)\Gamma(A).$$

Thus $K(t)$ has the limiting distribution $p(\infty)\Gamma$ as $t \to \infty$. Both $B(t)$ and $F(t)$ have the limiting distribution $p(\infty)\gamma$.

(x) *Kendall's non-occurrence formula.* For a bounded interval $I = (a, b)$, consider the probability Q_I that $Z(t) = 0$ for all $t \in I$. Then, for any $c \in I$,

$$Q_I = \mathbf{P}\{B(c) \geqslant c - a, F(c) \geqslant b - c\}$$
$$= \iint_{\substack{b \geqslant c-a \\ f \geqslant b-c}} p(c - b)\Gamma(db\,df)$$
$$= \int_0^a p(u)\mu[b - u, \infty]\,du.$$

This expression is, as it should be, independent of c, and we have the formula

$$(6) \qquad Q_I = \int_0^a p(u)m(b - u)\, du,$$

which is due to Kendall [3]. The length of the largest interval containing t in which $Z = 0$ is $L(t) = B(t) + F(t)$, and the reader will verify that L is not a Markov process.

(**xi**) The fact that B and F are Markov process, not independent of one another, such that (B, F) is a Markov process, raises two general questions.

(a) If X and Y are Markov processes, under what conditions is (X, Y) a Markov process?

(b) If (X, Y) is a Markov process, what conditions on its transition function suffice to ensure that X and Y are Markov processes?

(**xii**) *A result of Spitzer.* The argument used to prove Theorem 4.3 really establishes very much more. Let K be a Markov process on the non-negative quadrant with the property that, if $K(s) = (b, f)$, then

$$K(s + t) = (b + t, f - t)$$

for all $t \leqslant f$, and if $K(s) = (b, 0)$, the behaviour of $K(s + t)$ for $t > 0$ is the same as if $K(s) = (0, 0)$. Then the transition function of K must satisfy (3.16). The argument following (3.16) is one way (and there are others) of inserting the extra requirement that, from $(0, 0)$ the process jumps to a point $(0, f)$ rather than to a point in the interior of the quadrant.

The situation is much simpler in one dimension, and a direct proof of Theorem 4.4 may be given along the following lines. Suppose that F is a Markov process with the property that, if $F(s) = f$, then $F(s + t) = f - t$ for $0 \leqslant t \leqslant f$. Then the transition function of F must have the form

$$P_t(x, A) = \epsilon_{x-t}(A) \qquad (t < x),$$
$$= P_{t-x}(A) \qquad (t \geqslant x).$$

The function

$$p(t) = P_t(\{0\})$$

is a p-function; suppose it is standard. Then a simpler form of the argument leading to (3.16) concludes that P must be of the form (4.3). This result (under rather different conditions) is due to Spitzer (private communication).

(**xiii**) The reader may like to consider in what sense B can be said to be obtained from F by reversing the direction of time, and what light this fact throws on the complexities of the proof of Theorem 4.5.

(xiv) A curious fact about any standard p-function p (and one of which I know no application) follows from (3.10): that the function

(7)
$$r(t) = \int_0^t p(u)\, du$$

is subadditive. To see this, integrate (3.10) with respect to t over $(0, T)$, to get

$$
\begin{aligned}
r(s + T) - r(s) - p(s)r(T) &= \int_0^T dt \int_0^s \int_0^t p(s - u)p(t - v)\Gamma_0(du\, dv) \\
&= \int_0^s \int_0^T p(s - u)r(T - v)\Gamma_0(du\, dv) \\
&\leqslant r(T) \int_0^s \int_0^T p(s - u)\Gamma_0(du\, dv) \\
&= r(T) \int_0^s p(s - u)m(u)\, du \\
&= r(T)\{1 - p(s)\}.
\end{aligned}
$$

Replacing T by t and rearranging, we have

(8)
$$r(s + t) \leqslant r(s) + r(t).$$

(xv) *Fictitious regenerative phenomena.* The process defined by (1) is an example of one which undoubtedly has, in some sense, a regenerative nature, although it falls only trivially within the scope of this theory since

(9)
$$\mathbf{P}\{Z(t) = 1\} = 0$$

for all $t > 0$. It is possible to develop a theory which covers phenomena satisfying (9), and which exhibits some of the features of the present study. For instance, the processes B, F, K and V still appear, but V has no drift term, and is determined only up to changes of time scale. Kendall's construction still works, and there is an intimate connection with Lévy's theory of local time.

The theory of fictitious regenerative phenomena (or the intrinsic theory of local time) is still in its infancy, although an important contribution may be found in [25]. It is important to realise however that such a theory will, in several respects, be noticeably weaker than that developed here, essentially because of the absence of drift in V. For example, the arguments of Chapter 3 are almost all concerned with the extraction of analytical information from the presence of the additive term θ on the right hand side of the fundamental equation (3.13). (See also the remarks of Chung and Williams in the discussion of [42].)

CHAPTER 5

Quasi-Markov Chains

5.1 DEFINITIONS AND ELEMENTARY PROPERTIES

The example which motivates the definition of a regenerative phenomenon is the process formed by amalgamating all but one of the states of a Markov chain into a single composite state. For some purposes, however, this is too drastic, and it is necessary to allow two or more states to retain their distinct identities. For example, the analysis of the non-diagonal transition function p_{ij} ($i \neq j$) requires both the states i and j to remain intact. It is not difficult to see how the definitions of Chapter 2 should be extended to meet this requirement, and this chapter is devoted to the resulting extensions of the theory. For any positive integer N (which is the number of states to be held aloof from the amalgamation) we write

$$S_N = \{0, 1, 2, \ldots, N\}, \qquad S_N' = \{1, 2, \ldots, N\}.$$

Definition. A *quasi-Markov chain* of order N is a stochastic process $(Z(t);$ $t \geqslant 0)$ taking values in the set S_N and such that, whenever

$$0 = t_0 < t_1 < t_2 < \ldots < t_n$$

and $i_0, i_1, i_2, \ldots, i_n \in S_N'$, then

(1) $\quad \mathbf{P}\{Z(t_r) = i_r \ (1 \leqslant r \leqslant n)|Z(t_0) = i_0\} = \prod_{r=1}^{n} p_{i_{r-1}i_r}(t_r - t_{r-1}),$

where p_{ij} ($1 \leqslant i, j \leqslant N$) are functions of $t > 0$.

It is important to stress that (1) is only required to hold if $i_r \neq 0$; the state 0 is anomalous. (If (1) were required to hold even if $i_r = 0$, then Z would be a Markov chain on S_N.) Note that a quasi-Markov chain of order 1 (if $Z(0) = 1$) is a regenerative phenomenon.

As usual we shall write

(2) $\qquad\qquad\qquad \mathbf{P}_a = \mathbf{P}\{\cdot\,|Z(0) = a\}$

for $a \in S_N'$; it will not be necessary to contemplate the possibility that $Z(0) = 0$. It will become clear that, under $\mathbf{P}_a$, the finite-dimensional distributions of Z are determined by the functions p_{ij}, which are conveniently assembled into a matrix-valued function

(3) $$\mathfrak{p}(t) = (p_{ij}(t); i, j = 1, 2, \ldots, N)$$

called the *p-matrix* of Z.

Let X be a Markov chain on the countable state space S, with transition probabilities $p_{ij}^X(t)$. Select N distinct states $x_1, x_2, \ldots, x_N$, and define a function $\psi: S \to S_N$ by

(4) $$\psi(x_j) = j \qquad (1 \leqslant j \leqslant N)$$
$$\psi(x) = 0 \qquad (x \notin \{x_1, x_2, \ldots, x_N\}).$$

Then it is immediate that

(5) $$Z(t) = \psi\{X(t)\}$$

satisfies (1) with

(6) $$p_{ij}(t) = p_{x_i x_j}^X(t),$$

so that (5) defines a quasi-Markov chain whose p-matrix is a submatrix of the transition matrix of X. This remark is particularly important when $N = 2$.

Let Z be a quasi-Markov chain of order N with p-matrix $\mathfrak{p}(t)$, let $\alpha_1, \alpha_2, \ldots, \alpha_M$ ($M \leqslant N$) be distinct elements of S_N', and define $\chi: S_N \to S_M$ by

(7) $$\chi(\alpha_j) = j \qquad (1 \leqslant j \leqslant M),$$
$$\chi(\alpha) = 0 \qquad (\alpha \notin \{\alpha_1, \alpha_2, \ldots, \alpha_M\}).$$

Then

(8) $$\tilde{Z}(t) = \chi\{Z(t)\}$$

satisfies (1), and is therefore a quasi-Markov chain of order M, with p-matrix given by

(9) $$\tilde{p}_{ij}(t) = p_{\alpha_i \alpha_j}(t) \qquad (1 \leqslant i, j \leqslant M).$$

In particular, taking $M = 1$, we see that the process Z_j defined by

(10) $$Z_j(t) = 1 \qquad \text{if } Z(t) = j,$$
$$= 0 \qquad \text{otherwise,}$$

is a regenerative phenomenon under the probability measure $\mathbf{P}_j$, with p-function p_{jj}. Thus the diagonal elements of p-matrices are p-functions.

If $i \neq j$, then

(11) $\qquad \mathbf{P}_i\{Z_j(t_r) = 1 \ (1 \leqslant r \leqslant n)\} = p_{ij}(t_1) \prod_{r=2}^{n} p_{jj}(t_r - t_{r-1}),$

so that, under $\mathbf{P}_i$, Z_j is a delayed regenerative phenomenon with p-function p_{jj} and p^0-function p_{ij}. Thus, in the notation of §2.6,

(12) $\qquad\qquad\qquad\qquad p_{ij} \in \mathscr{L}(p_{jj}).$

The quasi-Markov chain (or the corresponding p-matrix) is said to be *standard* if each of its diagonal elements is a standard p-function. In view of the evident inequalities

(13) $\qquad\qquad\qquad p_{ij}(t) \geqslant 0, \quad \sum_{j=1}^{N} p_{ij}(t) \leqslant 1,$

this condition is equivalent to the assertion that

(14) $\qquad\qquad\qquad\qquad \lim_{t \to 0} \mathrm{p}(t) = I,$

where $I = I_N$ is the identity matrix of order N, and (here as elsewhere) the convergence of finite matrices is the convergence of their elements.

Theorem 5.1. *In order that an $(N \times N)$-matrix-valued function $\mathrm{p}(t)$ $(t > 0)$ should be the p-matrix of some quasi-Markov chain Z of order N, it is necessary and sufficient that, whenever $0 < t_1 < t_2 < \ldots < t_n$ and $1 \leqslant r \leqslant n$, the matrices*

(15) $\quad F(t_1, t_2, \ldots, t_r; \mathrm{p})$

$\qquad\qquad = \mathrm{p}(t_r) - \sum_{1 \leqslant \alpha < r} \mathrm{p}(t_\alpha)\mathrm{p}(t_r - t_\alpha)$

$\qquad\qquad + \sum_{1 \leqslant \alpha < \beta < r} \mathrm{p}(t_\alpha)\mathrm{p}(t_\beta - t_\alpha)\mathrm{p}(t_r - t_\beta)$

$\qquad\qquad - \ldots + (-1)^{r-1}\mathrm{p}(t_1)\mathrm{p}(t_2 - t_1) \ldots \mathrm{p}(t_r - t_{r-1})$

should satisfy

(16) $\quad F_{ij}(t_1, \ldots, t_r; \mathrm{p}) \geqslant 0, \quad \sum_{j=1}^{N} \sum_{r=1}^{n} F_{ij}(t_1, \ldots, t_r; \mathrm{p}) \leqslant 1,$

$(i, j \in S'_N)$. *The finite-dimensional distributions of Z, given $Z(0) \neq 0$, are determined by $\mathrm{p}(t)$.*

Proof. This follows very closely the argument used to prove Theorem 2.1 (which is the case $N = 1$ of this result) and will therefore be fairly briefly sketched; the details may be found in [39]. We first write (1) in the form

$$\mathbf{E}_{i_0} \prod_{r=1}^{n} Z_{i_r}(t_r) = \prod_{r=1}^{n} p_{i_{r-1}i_r}(t_r - t_{r-1}),$$

which shows that p determines the expectation under $\mathbf{P}_i$ $(i \neq 0)$ of any linear combination of products of the $Z_j(t)$ $(j \neq 0)$. Moreover,

$$Z_0(t) = 1 - \sum_{j=1}^{N} Z_j(t),$$

so that p determines

(17) $$\mathbf{E}_i \prod_{r=1}^{n} Z_{a_r}(t_r) = \mathbf{P}_i\{Z(t_r) = \alpha_r \ (1 \leqslant r \leqslant n)\}$$

even if some of the α_r are equal to 0. Thus the finite-dimensional distributions of Z under $\mathbf{P}_i$ $(i \neq 0)$ are determined by p.

In particular, p determines

(18) $$F_{ij}(t_1, \ldots, t_r; \mathfrak{p}) = \mathbf{P}\{Z(t_k) = 0 \ (1 \leqslant k < r), Z(t_r) = j\},$$

and expansion of the product (17) in this case shows that F_{ij} has the explicit expression (15). Thus the first inequality of (16) is necessary, as also is the second, since

$$\sum_{j=1}^{N} \sum_{r=1}^{n} F_{ij}(t_1, \ldots, t_r; \mathfrak{p}) = \sum_{r=1}^{n} \mathbf{P}\{Z(t_k) = 0 \ (1 \leqslant k < r), Z(t_r) \neq 0\}$$
$$= \mathbf{P}\{Z(t_r) \neq 0 \text{ for some } r \ (1 \leqslant r \leqslant n)\}.$$

Conversely, suppose that p satisfies (16). Then we may, as already indicated, use (1) to compute the probabilities (17), and the Daniell–Kolmogorov theorem will assure the existence of Z provided that these are non-negative. If $\alpha_r = 0$ for all $r \leqslant n - 1$ this is guaranteed by (16) (the first inequality if $\alpha_n \neq 0$, the second if $\alpha_n = 0$). On the other hand, if $\alpha_r \neq 0$ for some $r < n$, the argument already used in case $N = 1$ shows that (17) breaks up into a product of similar probabilities with smaller values of n. Induction on n completes the proof. ◆

Taking $n = 1$ in (16) recovers the inequalities (13). Less obvious are those when $n = 2$, which take the following form if we write $s = t_1$, $t = t_2 - t_1$ and **1** for the column vector whose components are all 1.

Corollary 1. *If* p *is a p-matrix, and* $s, t > 0$, *then*

(19) $$\mathfrak{p}(s + t) \geqslant \mathfrak{p}(s)\mathfrak{p}(t)$$

and

(20) $$\{I - \mathfrak{p}(s) - \mathfrak{p}(s + t) + \mathfrak{p}(s)\mathfrak{p}(t)\}\mathbf{1} \geqslant 0,$$

where the inequalities are to be interpreted elementwise.

If for the moment we write R for the non-negative matrix

$$R = \mathfrak{p}(s + t) - \mathfrak{p}(s)\mathfrak{p}(t),$$

then (20) implies that

$$\sum_{j=1}^{N} r_{ij} = (R\mathbf{1})_i \leqslant [\{I - \mathfrak{p}(s)\}\mathbf{1}]_i$$
$$= 1 - \sum_{k=1}^{N} p_{ik}(s).$$

Hence, for any $j \in S'_N$,

$$0 \leqslant r_{ij} \leqslant 1 - \sum_{k=1}^{N} p_{ik}(s),$$

so that

$$\sum_{k=1}^{N} p_{ik}(s)p_{kj}(t) \leqslant p_{ij}(s + t)$$
$$\leqslant \sum_{k=1}^{N} p_{ik}(s)p_{kj}(t) + 1 - \sum_{k=1}^{N} p_{ik}(s)$$
$$= 1 - \sum_{k=1}^{N} p_{ik}(s)\{1 - p_{kj}(t)\}.$$

Thus

$$p_{ii}(s)p_{ij}(t) \leqslant p_{ij}(s + t)$$
$$\leqslant 1 - p_{ii}(s)\{1 - p_{ij}(t)\},$$

or equivalently

$$-\{1 - p_{ii}(s)\}p_{ij}(t) \leqslant p_{ij}(s + t) - p_{ij}(t)$$
$$\leqslant \{1 - p_{ii}(s)\}\{1 - p_{ij}(t)\},$$

which shows that

(21) $$|p_{ij}(s + t) - p_{ij}(t)| \leqslant 1 - p_{ii}(s).$$

In particular, when $\mathfrak{p}$ is standard we have the following corollary.

Corollary 2. *Every element p_{ij} of a standard p-matrix $\mathfrak{p}$ is uniformly continuous on $(0, \infty)$.*

Already therefore we know a good deal about the functions p_{ij} which go to make up a standard p-matrix $\mathfrak{p}$. The diagonal functions $(i = j)$ are standard p-functions, and therefore have all the properties established in Chapters 2 and 3 for members of $\mathcal{P}$. The non-diagonal functions are continuous, and belong to the respective classes $\mathcal{L}(p_{jj})$. Thus by Theorem 2.8 there is a probability measure ϕ_{ij} on $[0, \infty]$ such that

$$(22) \qquad p_{ij}(t) = \int_0^t p_{jj}(t - s)\phi_{ij}(ds).$$

Since p_{ij} is continuous, and tends to 0 as $t \to 0$, ϕ_{ij} can have no atoms except perhaps at ∞. It follows at once from (22) that p_{ij} has bounded variation on every finite interval (since p_{jj} does), so that p_{ij} is differentiable almost everywhere, and that the limit

$$(23) \qquad p_{ij}(\infty) = \lim_{t \to \infty} p_{ij}(t)$$

exists; indeed

$$(24) \qquad p_{ij}(\infty) = p_{jj}(\infty)\phi_{ij}[0, \infty).$$

In order to go further, we need some criterion for p to be a p-matrix which is more usable than that of Theorem 5.1. The key tool is a generalisation of Theorem 3.1 which will be proved in the next section.

5.2 THE CHARACTERISATION THEOREM

If $\mathfrak{f}(t) = (f_{ij}(t); i, j = 1, 2, \ldots, N)$ is any matrix-valued function of $t > 0$, we shall denote by $\hat{\mathfrak{f}}(\theta)$ its elementwise Laplace transform, the matrix whose elements are

$$\hat{f}_{ij}(\theta) = \int_0^\infty f_{ij}(t)\, e^{-\theta t}\, dt.$$

Theorem 5.2. *If* $\mathfrak{p}(t)$ *is a standard p-matrix, then its Laplace transform may be written, for* $\theta > 0$, *in the form*

$$(1) \qquad \hat{\mathfrak{p}}(\theta) = \mathfrak{q}(\theta)^{-1},$$

where the elements $q_{ij}(\theta)$ *of* $\mathfrak{q}(\theta)$ *are of the form*

$$(2) \qquad q_{ii}(\theta) = \theta + \int(1 - e^{-\theta x})\mu_i(dx),$$

$$(3) \qquad q_{ij}(\theta) = -\int e^{-\theta x}\lambda_{ij}(dx) \qquad (i \neq j),$$

μ_i *is a measure* on* $(0, \infty]$ *satisfying*

(4) $$\int (1 - e^{-x})\mu_i(dx) < \infty,$$

λ_{ij} $(i \neq j)$ *is a totally finite measure on* $[0, \infty)$, *and for each i,*

(5) $$\sum_{j=1}^{N} \lambda_{ij}[0, \infty) \leqslant \mu_i\{\infty\}.$$

These measures are uniquely determined by $\mathfrak{p}$. *Conversely, if for* $i, j = 1, 2,$
$\ldots, N$, μ_i *and* λ_{ij} $(i \neq j)$ *are measures on* $(0, \infty]$ *and* $[0, \infty)$ *respectively,*
satisfying (4) *and* (5), *then there is a unique continuous matrix-valued*
function $\mathfrak{p}(t)$ *satisfying* (1), (2) *and* (3), *and* $\mathfrak{p}$ *is a standard p-matrix.*

Proof. Let $\mathfrak{p}$ be a standard p-matrix, and write

$$f_{ij}(n, h) = F_{ij}(h, 2h, \ldots, nh; \mathfrak{p})$$

for $n \geqslant 1, h > 0$. Then

$$p_{ij}(nh) = \mathbf{P}_i\{Z(nh) = j\}$$

$$= \sum_{r=1}^{n} \mathbf{P}_i\{Z(sh) = 0 \ (1 \leqslant s < r), Z(rh) \neq 0, Z(nh) = j\}$$

$$= \sum_{r=1}^{n} \sum_{k=1}^{N} f_{ik}(r, h)p_{kj}(nh - rh),$$

or in matrix notation (and using the obvious convention that $\mathfrak{p}(0) = I$),

$$\mathfrak{p}(nh) = \sum_{r=1}^{n} \mathfrak{f}(r, h)\mathfrak{p}[(n - r)h].$$

Hence, for $|z| < 1$,

$$\sum_{n=0}^{\infty} \mathfrak{p}(nh)z^n = I + \sum_{r=1}^{\infty} \mathfrak{f}(r, h)z^r \sum_{n=0}^{\infty} \mathfrak{p}(nh)z^n,$$

or

$$\sum_{n=0}^{\infty} \mathfrak{p}(nh)z^n = \left\{ I - \sum_{r=1}^{\infty} \mathfrak{f}(r, h)z^r \right\}^{-1}.$$

---

* The reader is reminded of the convention that all measures defined on intervals of the
real line are positive Borel measures, and that the range of integration is omitted in
integrals like (2) and (3) which are over the whole interval of definition.

Fixing a positive θ, and writing $z = e^{-\theta h}$, the continuity and boundedness of the p_{ij} yields (as on page 58) the formula for $\hat{p}(\theta)$:

$$\hat{p}(\theta) = \lim_{h \to 0} h \sum_{n=0}^{\infty} p(nh) \, e^{-\theta nh}$$

$$= \lim_{h \to \infty} \left\{ \frac{I - \sum_{r=1}^{\infty} f(r, h) \, e^{-\theta rh}}{h} \right\}^{-1}.$$

Denote by $\mathscr{Z}$ the set

$$\mathscr{Z} = \{\theta > 0; \det \hat{p}(\theta) = 0\}.$$

Since by (1.14),

(6)
$$\lim_{\theta \to \infty} \theta \hat{p}(\theta) = I,$$

we have

$$\lim_{\theta \to \infty} \theta^N \det \hat{p}(\theta) = 1,$$

so that $\mathscr{Z}$ is bounded. Since $\det \hat{p}(\theta)$ is analytic in $\mathrm{Re}\ \theta > 0$, $\mathscr{Z}$ can have no limit point in $(0, \infty)$, though it might perhaps have 0 as a limit point.

For $\theta \notin \mathscr{Z}$, $q(\theta) = \hat{p}(\theta)^{-1}$ exists, and

$$q(\theta) = \lim_{h \to 0} h^{-1} \left[I - \sum_{r=1}^{\infty} f(r, h) z^r \right].$$

We have to examine the elements of this matrix equation separately in the diagonal and non-diagonal cases.

For any i, and $\theta \notin \mathscr{Z}$,

$$q_{ii}(\theta) = \lim_{h \to 0} h^{-1} \left\{ 1 - \sum_{r=1}^{\infty} f_{ii}(r, h) \, e^{-\theta rh} \right\}.$$

To this equation we apply word for word the argument of §3.2, using the facts that

$$\sum_{r=1}^{\infty} f_{ii}(r, h) \leqslant 1$$

and that (6) implies that

$$\lim_{\theta \to \infty} \theta^{-1} q_{ii}(\theta) = 1,$$

to establish the existence of a measure μ_i satisfying (4) and such that, for $\theta \notin \mathscr{Z}$,

$$q_{ii}(\theta) = \theta + \int (1 - e^{-\theta x}) \mu_i(dx).$$

The only complication is that θ must avoid the (at most countable) set $\mathscr{L}$. This means that we have to show that, for finite measures λ and λ' on $[0, \infty)$,

$$\int e^{-\theta x} \lambda(dx) = \int e^{-\theta x} \lambda'(dx)$$

for all $\theta > 0$, $\theta \notin \mathscr{L}$ implies $\lambda = \lambda'$, and this is certainly true since the complement of $\mathscr{L}$ is dense.

For $i \neq j$, $\theta \notin \mathscr{L}$,

$$q_{ij}(\theta) = -\lim_{h \to 0} h^{-1} \sum_{r=1}^{\infty} f_{ij}(r, h)\, e^{-\theta rh}$$

$$= -\int e^{-\theta x} \lambda_{ij}^{(h)}(dx),$$

where $\lambda_{ij}^{(h)}$ is the measure on $[0, \infty)$ which assigns mass $h^{-1} f_{ij}(r, h)$ to the point $x = rh$. Weak convergence arguments (of which the details are not quite trivial, and are given in the Appendix) then establish the existence of a measure λ_{ij} on $[0, \infty)$ such that

$$q_{ij}(\theta) = -\int e^{-\theta x} \lambda_{ij}(dx)$$

for all $\theta > 0$, $\theta \notin \mathscr{L}$. (It is not at this stage asserted that λ_{ij} is totally finite.)

Thus we have proved that (2) and (3) hold, except for θ in the exceptional set $\mathscr{L}$. However, since $\mathscr{L}$ is countable, its complement is dense in $(0, \infty)$, so that the integrals in (2) and (3) converge, and define continuous functions of θ, in $\theta > 0$. Hence there is a continuous function $q(\theta)$ in $\theta > 0$ such that

$$\hat{p}(\theta) q(\theta) = I$$

for $\theta \notin \mathscr{L}$. By continuity this equation holds for all $\theta > 0$, so that $\hat{p}(\theta)$ is always non-singular, and $\mathscr{L}$ is empty.

To prove (5), note that

$$\sum_{j=1}^{N} q_{ij}(\theta) = \lim_{h \to 0} h^{-1} \left\{ 1 - \sum_{j=1}^{N} \sum_{r=1}^{\infty} f_{ij}(r, h)\, e^{-\theta rh} \right\}$$

$$\geqslant \lim_{h \to 0} h^{-1} \left\{ 1 - \sum_{j=1}^{N} \sum_{r=1}^{\infty} f_{ij}(r, h) \right\} \geqslant 0$$

by (1.16). Hence, for each i,

$$\sum_{j \neq i} \int e^{-\theta x} \lambda_{ij}(dx) \leqslant \theta + \int (1 - e^{-\theta x}) \mu_i(dx),$$

whence (5) follows on letting $\theta \to 0$. Notice that this now implies that $\lambda_{ij}[0, \infty) < \infty$.

The fact that μ_i and λ_{ij} are determined uniquely by p follows of course from (1), (2) and (3) and the uniqueness theorem for Laplace transforms. Hence the first half of the theorem is proved.

Conversely, suppose that measures μ_i, λ_{ij} are given satisfying (4) and (5), and define q(θ) by (2) and (3). We shall assume that there is no value of i for which all the measures λ_{ij} vanish identically, leaving the reader to supply the simple modifications needed to cover this case. Then, by (5), $\mu_i\{\infty\} > 0$, and we shall write

$$
\begin{aligned}
(7) \qquad a_{ij} &= \lambda_{ij}[0, \infty)/\mu_i\{\infty\} \qquad (i \neq j) \\
&= 0 \qquad\qquad\qquad\quad (i = j),
\end{aligned}
$$

so that $A = (a_{ij})$ is a substochastic matrix. Then we construct a discrete-time Markov chain $X = (X_0, X_1, \ldots)$ with state space S_N' and transition matrix A.

By Theorem 3.1, there is a standard p-function p_i with

$$
(8) \qquad \hat{p}_i(\theta) = \left\{ \theta + \int (1 - e^{-\theta x})\mu_i(dx) \right\}^{-1} = q_{ii}(\theta)^{-1},
$$

and since $\mu_i\{\infty\} > 0$ this is transient. Given X, we may therefore construct independent regenerative phenomena $Z_0, Z_1, Z_2, \ldots$, where if $X_n = i$, Z_n has p-function p_i. Since Z_n is transient, it has a finite lifetime

$$
\zeta_n = \sup \{t; Z_n(t) = 1\}.
$$

Given X, construct random variables $\eta_0, \eta_1, \eta_2, \ldots$, independent of one another and of the Z_n, such that if $X_n = i$ and $X_{n+1} = j$, then the distribution of η_n is given by the probability measure

$$
\lambda_{ij}/\lambda_{ij}[0, \infty),
$$

and $\eta_n = 0$ if $\lambda_{ij}[0, \infty) = 0$. Write

$$
\xi_n = \sum_{r=0}^{n-1} (\zeta_r + \eta_r),
$$

and define a process Z by

$$
\begin{aligned}
(9) \quad Z(t) = i \quad &\text{if, for some } n, \ t \in [\xi_n, \xi_n + \zeta_n], \\
&\qquad X_n = i, Z_n(t - \xi_n) = 1, \\
= 0 \quad &\text{otherwise.}
\end{aligned}
$$

The starting point $Z(0)$ can be an arbitrary point X_0 of S_N'. It is asserted that Z is a standard quasi-Markov chain, and that its p-matrix satisfies (1).

To show that Z is a quasi-Markov chain, we have only to note that, for any $T > 0$, $i \in S'_N$, the event $\{Z(T) = i\}$ entails that, for some n,

$$\xi_n \leqslant T < \xi_n + \zeta_n, \qquad X_n = i, \qquad Z_n(t - \xi_n) = 1.$$

By the regenerative property of Z_n, the Markovian nature of X, and the independence of the η_n, the conditional distributions of $(Z(T + t); t > 0)$ are the same as those of $(Z(t); t > 0)$ conditional on $Z(0) = i$. Thus

$$\mathbf{P}\{Z(T + t_r) = i_r \, (1 \leqslant r \leqslant n)|Z(T) = i\}$$
$$= \mathbf{P}_i\{Z(t_r) = i_r \, (1 \leqslant r \leqslant n)\},$$

which is enough to show that Z is a quasi-Markov chain.

Let p be the p-matrix of Z. Then

$$p_{ij}(t) = \sum_{n=0}^{\infty} \mathbf{P}\{\xi_n \leqslant t < \xi_n + \zeta_n, X_n = j, Z_n(t - \xi_n) = 1 | X_0 = i\}$$
$$= \sum_{n=0}^{\infty} \mathbf{E}\{p_j(t - \xi_n); X_n = j | X_0 = i\}.$$

In particular,

$$p_{ii}(t) \geqslant p_i(t) \to 1$$

as $t \to 0$, so that p is standard. Thus the functions p_{ij} are continuous, and so determined by their Laplace transforms (for $\theta > 0$)

$$\hat{p}_{ij}(\theta) = \sum_{n=0}^{\infty} \int_0^{\infty} \mathbf{E}\{p_j(t - \xi_n); X_n = j | X_0 = i\} \, \mathrm{e}^{-\theta t} \, dt$$
$$= \sum_{n=0}^{\infty} \mathbf{E}\{\mathrm{e}^{-\theta \xi_n} \hat{p}_j(\theta); X_n = j | X_0 = i\}$$
$$= \hat{p}_j(\theta) \sum_{n=0}^{\infty} \mathbf{E}\{\mathrm{e}^{-\theta \xi_n}; X_n = j | X_0 = i\}.$$

Now

$$\mathbf{E}\{\mathrm{e}^{-\theta \xi_n}; X_n = j | X_0 = i\}$$
$$= \sum_{k_1, k_2, \ldots, k_{n-1} = 1}^{N} a_{ik_1} a_{k_1 k_2} \cdots a_{k_{n-1} j} \prod_{r=0}^{n-1} \mu_{k_r}\{\infty\} \hat{p}_{k_r}(\theta) \frac{\hat{\lambda}_{k_r k_{r+1}}(\theta)}{\lambda_{k_r k_{r+1}}[0, \infty)}$$
$$= \sum_{k_1, k_2, \ldots, k_{n-1} = 1}^{N} b_{ik_1} b_{k_1 k_2} \cdots b_{k_{n-1} j},$$

where

$$b_{ij} = a_{ij} \mu_i\{\infty\} \hat{p}_i(\theta) \hat{\lambda}_{ij}(\theta)/\lambda_{ij}[0, \infty)$$
$$= \hat{\lambda}_{ij}(\theta) \hat{p}_i(\theta),$$

and

$$\hat{\lambda}_{ij}(\theta) = \int e^{-\theta x} \lambda_{ij}(dx).$$

Thus, in matrix notation,

$$\mathfrak{p}(\theta) = \sum_{n=0}^{\infty} B^n \, \text{diag} \, [\hat{p}_j(\theta)]$$
$$= (I - B)^{-1} \, \text{diag} \, [\hat{p}_j(\theta)],$$

so that

$$\mathfrak{p}(\theta) = \bar{\mathfrak{q}}(\theta)^{-1},$$

where

$$\bar{\mathfrak{q}}(\theta) = \text{diag} \, [\hat{p}_j(\theta)^{-1}](I - B).$$

For $i \neq j$,

$$\bar{q}_{ij}(\theta) = -\hat{p}_i(\theta)^{-1} b_{ij} = -\hat{\lambda}_{ij}(\theta) = q_{ij}(\theta),$$

and

$$\bar{q}_{ii}(\theta) = \hat{p}_i(\theta)^{-1} = q_{ii}(\theta).$$

Hence

$$\bar{\mathfrak{q}}(\theta) = \mathfrak{q}(\theta),$$

and the proof is complete.   ◆

This construction makes clear the meaning of the measures μ_i and λ_{ij} which appear in the canonical representation (1) of $\hat{\mathfrak{p}}(\theta)$. The measure μ_i is the canonical measure of the p-function p_i. If $Z(0) = i$, p_i is the p-function of the regenerative phenomenon Z_0 with which Z coincides (except for labelling) up to the time when Z first enters a state other than i or 0. Thus p_i can be regarded as a taboo probability in the sense of Chung [6]; for the version of Z constructed in the proof of the theorem (and therefore for any separable quasi-Markov chain with the given p-matrix, by the arguments of §4.1),

(10) $p_i(t) = \mathbf{P}_i\{Z(s) \in \{0, i\} \ (0 < s < t), Z(t) = i\}.$

Similarly, λ_{ij} has the following meaning. Let Z be separable, $Z(0) = i$, and write

$$\tau = \inf \{t > 0; Z(t) = j\},$$
$$\sigma = \sup \{t < \tau; Z(t) = i\}.$$

Thus λ_{ij} is a multiple of the distribution, conditional on $\{\tau < \infty\}$, of $(\tau - \sigma)$.

The total mass of λ_{ij} plays its part in establishing the transition matrix A, which describes the movement of Z among the states in S'_N, if time in 0

is ignored. Notice that A may be substochastic, so that (X_n) may have a finite lifetime. This means that the process Z may ultimately remain for ever in 0.

5.3 THE DECOMPOSITIONS

The most important applications of Theorem 5.2 are in the case $N = 2$, for which stronger conclusions flow from the fact that the inversion of a (2×2) matrix is particularly simple. In fact, if as before we write $\hat{\lambda}_{ij}$ for the Laplace–Stieltjes transform of the measure λ_{ij}, and p_i for the p-function with canonical measure μ_i, so that

(1)
$$\hat{p}_i(\theta) = \left\{ \theta + \int (1 - e^{-\theta x}) \mu_i(dx) \right\}^{-1},$$

then (2.1) takes the form

$$\mathfrak{p}(\theta) = \begin{pmatrix} \hat{p}_1(\theta)^{-1} & -\hat{\lambda}_{12}(\theta) \\ -\hat{\lambda}_{21}(\theta) & \hat{p}_2(\theta)^{-1} \end{pmatrix}^{-1}$$
$$= \Delta(\theta) \begin{pmatrix} \hat{p}_2(\theta)^{-1} & \hat{\lambda}_{12}(\theta) \\ \hat{\lambda}_{21}(\theta) & \hat{p}_1(\theta)^{-1} \end{pmatrix},$$

where

(2)
$$\Delta(\theta) = \det \hat{\mathfrak{p}}(\theta).$$

Hence

(3)
$$\hat{p}_{11}(\theta) = \Delta(\theta)/\hat{p}_2(\theta),$$
$$\hat{p}_{12}(\theta) = \Delta(\theta)\hat{\lambda}_{12}(\theta),$$
$$\hat{p}_{21}(\theta) = \Delta(\theta)\hat{\lambda}_{21}(\theta),$$
$$\hat{p}_{22}(\theta) = \Delta(\theta)/\hat{p}_1(\theta).$$

These formulae imply a considerable improvement to the *first passage decomposition* (1.22), and yield a dual *last exit decomposition*.

Theorem 5.3. *If p is a standard p-matrix of order 2, then*

(4)
$$p_{12}(t) = \int_0^t f_{12}(s) p_{22}(t - s) \, ds,$$

where

(5)
$$f_{12}(t) = \int_0^t p_1(t - u) \lambda_{12}(du)$$

is a non-negative integrable function with

(6)
$$\int_0^\infty f_{12}(t)\, dt \leqslant 1.$$

Moreover,

(7)
$$p_{12}(t) = \int_0^t p_{11}(s) g_{12}(t - s)\, ds,$$

where

(8)
$$g_{12}(t) = \int_0^t p_2(t - u)\lambda_{12}(du).$$

Proof. From (3),

$$\hat{p}_{12}(\theta)/\hat{p}_{22}(\theta) = \hat{p}_1(\theta)\hat{\lambda}_{12}(\theta) = \hat{f}_{12}(\theta),$$

where f_{12} is given by (5). Hence

$$\hat{p}_{12}(\theta) = \hat{f}_{12}(\theta)\hat{p}_{22}(\theta),$$

and, since p_{12} is continuous, (4) follows. The measure ϕ_{12} of (1.22) therefore has density f_{12} in $(0, \infty)$, so that

$$\int_0^\infty f_{12}(t)\, dt = \phi_{12}(0, \infty) \leqslant \phi_{12}(0, \infty] = 1.$$

The proof of (7) is similar, since

$$\hat{p}_{12}(\theta)/\hat{p}_{11}(\theta) = \hat{p}_2(\theta)\hat{\lambda}_{12}(\theta) = \hat{g}_{12}(\theta). \qquad \blacklozenge$$

It is convenient to write the above equations in convolution notation as

(4′) $p_{12} = f_{12} * p_{22},$

(5′) $f_{12} = p_1 * d\lambda_{12},$

(7′) $p_{12} = p_{11} * g_{12},$

(8′) $g_{12} = p_2 * d\lambda_{12},$

where the convolution of two functions is

$$(\phi * \psi)(t) = \int_0^t \phi(s)\psi(t - s)\, ds,$$

and that of a function and a measure is

$$(\phi * d\mu)(t) = \int_0^t \phi(t - s)\mu(ds).$$

Such convolutions are associative, so that (4′) and (5′) can be combined to give

(9) $$p_{12} = p_1 * d\lambda_{12} * p_{22},$$

and (7′) and (8′) as

(10) $$p_{12} = p_{11} * d\lambda_{12} * p_2.$$

These expressions for the non-diagonal elements of a p-matrix are characteristic, in a sense made precise by the following theorem.

Theorem 5.4. *Let $N \geqslant 2$ be a fixed integer, and f a function on $(0, \infty)$. In order that there should exist a standard p-matrix having f as a diagonal element, it is necessary and sufficient that f belong to $\mathscr{P}$. On the other hand, in order that there should exist a standard p-matrix with f as a non-diagonal element, it is necessary and sufficient that f be expressible in the form*

(11) $$f = \acute{p} * d\lambda * \grave{p},$$

where $\acute{p}$ and $\grave{p}$ belong to $\mathscr{P}$, λ is a totally finite measure on $[0, \infty)$, and

(12) $$\int_0^\infty \acute{p}(t)\, dt\; \lambda[0, \infty) \leqslant 1.$$

Proof—diagonal case. We have already seen that, if p is a standard p-matrix, then $p_{ii} \in \mathscr{P}$ for each i. Conversely, if $f \in \mathscr{P}$, and μ is the canonical measure of f, let p be the p-matrix defined by (2.1), with

$$\mu_1 = \mu, \qquad \mu_i = 0 \; (i \geqslant 2), \qquad \lambda_{ij} = 0 \; (i, j = 1, 2, \ldots, N).$$

Then an easy calculation yields

$$p(t) = \text{diag}\,[f(t), 1, 1, \ldots, 1],$$

so that f is a diagonal element of p.

Non-diagonal case. Suppose that p is a standard p-matrix of order N, and that $f = p_{ij}$ for a pair of distinct integers i, j. It was remarked in §5.1 that any principal submatrix of a p-matrix is a p-matrix, so that

$$\begin{pmatrix} p_{ij} & p_{ij} \\ p_{ji} & p_{ji} \end{pmatrix}$$

is a standard p-matrix of order 2. Applying Theorem 5.3 to this p-matrix establishes the necessity of (11) and (12).

Conversely, suppose that f has the form (11), and let $\acute{\mu}$ and $\grave{\mu}$ be the canonical measures of $\acute{p}$ and $\grave{p}$. Define a standard p-matrix of order N by (2.1), where

$$\mu_1 = \acute{\mu}, \qquad \mu_2 = \grave{\mu}, \qquad \mu_3 = \mu_4 = \ldots = \mu_N = 0,$$
$$\lambda_{12} = \lambda, \qquad \lambda_{ij} = 0, \qquad ((i, j) \neq (1, 2)).$$

It is easy to check that (2.4) and (2.5) are satisfied, and that

$$\hat{p}(\theta) = \begin{pmatrix} \acute{p}(\theta)^{-1} & -\lambda(\theta) & 0 \ldots 0 \\ 0 & \grave{p}(\theta)^{-1} & 0 \ldots 0 \\ 0 & 0 & \theta \ldots 0 \\ \cdot & \cdot & \cdot \\ 0 & 0 & 0 \ldots \theta \end{pmatrix}^{-1}$$

$$= \begin{pmatrix} \acute{p}(\theta) & \acute{p}(\theta)\hat{\lambda}(\theta)\grave{p}(\theta) & 0 & \ldots & 0 \\ 0 & \grave{p}(\theta) & 0 & \ldots & 0 \\ 0 & 0 & \theta^{-1} & \ldots & 0 \\ \cdot & \cdot & \cdot & & \cdot \\ 0 & 0 & 0 & \ldots & \theta^{-1} \end{pmatrix},$$

so that

$$\hat{p}_{12}(\theta) = \acute{p}(\theta)\hat{\lambda}(\theta)\grave{p}(\theta) = f(\theta),$$

and

$$p_{12} = f. \qquad \blacklozenge$$

A simple, but very important, corollary of this result deserves the status of a theorem.

Theorem 5.5. *If* p *is a standard p-matrix, then every non-diagonal element* p_{ij} *($i \neq j$) has a finite derivative at the origin*

$$(13) \qquad\qquad q_{ij} = \lim_{t \to 0} t^{-1} p_{ij}(t).$$

Proof. By the last theorem, it suffices to prove that every function f of the form (11) has finite derivative $f'(0)$. To do this, note that the function

$$\delta(t) = \sup_{0 \leqslant u \leqslant t} \{1 - \min[\acute{p}(u), \grave{p}(u)]\}$$

is non-decreasing, and that

$$\lim_{t \to 0} \delta(t) = 0.$$

Hence

$$(\dot{p} * \dot{p})(t) = \int_0^t \dot{p}(s)\dot{p}(t - s) \, ds$$

satisfies

$$t \geqslant (\dot{p} * \dot{p})(t) \geqslant \int_0^t [1 - \delta(t)]^2 \, ds$$
$$\geqslant [1 - 2\delta(t)]t,$$

and so

$$\int_0^t (t - u)\lambda(du) \geqslant f(t) \geqslant [1 - 2\delta(t)] \int_0^t (t - u)\lambda(du).$$

Since

$$\lim_{t \to \infty} \int_0^t \frac{t - u}{t} \lambda(du) = \lambda\{0\},$$

it follows that

$$\lim_{t \to 0} t^{-1}f(t) = \lambda\{0\}. \qquad \blacklozenge$$

Notice that, when $N = 2$, we have the explicit formula for q_{12}:

(14) $$q_{12} = \lambda_{12}\{0\}.$$

If we apply Theorem 5.5 to the quasi-Markov chain (1.5), we obtain the celebrated theorem of Kolmogorov that, in a Markov chain, p_{ij} has a finite derivative q_{ij} at the origin when $i \neq j$. The usual proofs of this result are quite different, and it is interesting to see how naturally it falls out from the decomposition formula (9).

5.4 THE ERGODIC LIMIT

Let p be a standard p-matrix of order N. From the relatively crude theory of §2.6 we already know that

(1) $$p_{ij}(\infty) = \lim_{t \to \infty} p_{ij}(t)$$

exists for all i, j. It is therefore natural to ask what can be said about the limit matrix

(2) $$\mathfrak{p}(\infty) = (p_{ij}(\infty)),$$

5

and in particular how it relates to the quantities appearing in the representation of Theorem 5.2.

As in Markov chain theory, the situation is most clear cut when p is *irreducible*, in the sense that none of the functions p_{ij} is identically zero. We first need a criterion for irreducibility.

Lemma. *If $N \geqslant 2$, the standard p-matrix* p *is irreducible if and only if $\mu_i\{\infty\} > 0$ for all i, and the Markov chain (on the finite state space S'_N) with transition probabilities*

$$(3) \qquad\qquad a_{ij} = \lambda_{ij}[0, \infty)/\mu_i\{\infty\}, \qquad a_{ii} = 0$$

is irreducible.

Proof. Suppose first that, for some i, $\mu_i\{\infty\} > 0$. Then $\lambda_{ij}[0, \infty) = 0$ for all j, and the ith row of $\hat{p}(\theta)^{-1}$ contains, by (2.1), only one non-zero term (in the diagonal place). Thus, by the usual rules for inverting partitioned matrices, the ith row of $\hat{p}(\theta)$ contains only one non-zero term, so that $p_{ij} = 0$ for all $j \neq i$, and p is not irreducible.

Secondly, suppose that $\mu_i\{\infty\} > 0$ for all i, so that (by (2.5)) (3) defines a substochastic matrix A, and suppose that A is not irreducible. Then there exists a proper subset C of S'_N such that

$$a_{ij} = 0 \qquad (i \notin C, j \in C).$$

Hence

$$\lambda_{ij} = 0 \qquad (i \notin C, j \in C),$$

so that

$$q_{ij}(\theta) = 0 \qquad (i \notin C, j \in C).$$

Again the use of partitioned matrices shows that

$$\hat{p}_{ij}(\theta) = 0 \qquad (i \notin C, j \in C),$$

and contradicts the irreducibility of p.

Thus we have proved that the conditions of the lemma are necessary for p to be irreducible. To prove that they are also sufficient, assume that they hold, but that p is not irreducible. Then there are integers $a, b \in S'_N$ such that $p_{ab}(t) = 0$ for all t, and

$$C = \{j \in S'_N; p_{aj}(t) = 0 \text{ for all } t\}$$

is a proper subset of S'_N since $a \notin C$, $b \in C$. For $i \notin C$, $j \in C$, use (1.19) to give, for $s, t > 0$

$$0 = p_{aj}(s + t)$$

$$\geqslant \sum_{k=1}^{N} p_{ak}(s)p_{kj}(t)$$

$$\geqslant p_{ai}(s)p_{ij}(t).$$

Since there are values of s with $p_{ai}(s) > 0$, we have

$$p_{ij}(t) = 0$$

for all $t > 0$, $i \notin C$, $j \in C$. Hence as before we deduce that, for $i \notin C$, $j \notin C$,

$$\hat{p}_{ij}(\theta) = 0,$$
$$q_{ij}(\theta) = 0,$$
$$\lambda_{ij} = 0,$$
$$a_{ij} = 0,$$

so that A is not irreducible. The contradiction completes the proof. ◆

Theorem 5.6. *If* p *is an irreducible, standard p-matrix, then*

(4) $$p_i = \lim_{t \to \infty} p_{ij}(\infty)$$

does not depend on i. The p_i are all zero unless there is equality in (2.5) for all i (i.e. A is stochastic) and

(5) $$m_i = \int_{(0,\infty)} x\mu_i(dx) < \infty,$$

(6) $$l_{ij} = \int_{(0,\infty)} x\lambda_{ij}(dx) < \infty,$$

for all i, j. If these conditions are satisfied, then the p_j are all positive, and are determined uniquely up to a constant factor by the equations

(7) $$p_j\mu_j\{\infty\} = \sum_{i \neq j} p_i\lambda_{ij}[0, \infty).$$

Proof. For $i, j, I, J \in S'_N$, (1.19) shows that

$$p_{ij}(t + a + b) \geqslant p_{iI}(a)p_{IJ}(t)p_{Jj}(b),$$

and by irreducibility we may choose a and b so that

$$c = p_{iI}(a)p_{Jj}(b) > 0.$$

Then

$$p_{ij}(\infty) \geqslant c p_{IJ}(\infty),$$

and therefore $p_{IJ}(\infty) > 0$ implies that $p_{ij}(\infty) > 0$. Thus the numbers $p_{ij}(\infty)$ are either all zero or all positive.

Suppose that $p_{ij}(\infty) > 0$ for some, and then for all, pairs i, j. Then

$$\mathfrak{p}(\infty) = \lim_{\theta \to 0} \theta \hat{\mathfrak{p}}(\theta),$$

so that

$$
\begin{aligned}
0 &= \lim_{\theta \to 0} \theta I \\
&= \lim_{\theta \to 0} \theta \hat{\mathfrak{p}}(\theta) \mathfrak{q}(\theta) \\
&= \mathfrak{p}(\infty) \mathfrak{q}(0).
\end{aligned}
$$

Thus

(8) $$\mathfrak{p}(\infty)\mathfrak{q}(0) = 0,$$

and similarly

(9) $$\mathfrak{q}(0)\mathfrak{p}(\infty) = 0.$$

Summing the rows of (8) we have

$$\sum_{k=1}^{N} p_{ik}(\infty)\{\mu_k\{\infty\} - \sum_{j \neq k} \lambda_{kj}[0, \infty)\} = 0,$$

which shows that equality obtains in (2.5). Thus A is an irreducible stochastic matrix, and it follows that the equations

(10) $$\zeta = \zeta A, \qquad z = Az$$

each have one and, up to constant multiples, only one solution. The solutions of the second equation are the scalar multiples of **1**. Now (9) shows that each column of $\mathfrak{p}(\infty)$ satisfies $z = Az$, so that each column is constant; $p_{ij}(\infty)$ does not depend on i.

Equation (8) now takes the form (7), which shows that

$$\zeta_j = p_j \mu_j\{\infty\}$$

satisfies $\zeta = \zeta A$, and determines the p_j up to a constant factor. It therefore remains only to show that the p_j are non-zero if and only if the integrals in (5) and (6) converge.

Suppose first that (5) and (6) are satisfied, and that there is equality in (2.5). Then it follows from (2.2) and (2.3) that $q_{ij}(\theta)$ has a finite (right) derivative at $\theta = 0$, so that the finite limit

$$\mathfrak{s} = \lim_{\theta \to 0} \theta^{-1}\{q(\theta) - q(0)\}$$

exists. Since

$$p(\infty) = \lim_{\theta \to 0} \theta\hat{p}(\theta),$$

we have

$$p(\infty)\mathfrak{s} = \lim_{\theta \to 0} \hat{p}(\theta)\{q(\theta) - q(0)\}$$

$$= I - \lim_{\theta \to 0} \hat{p}(\theta)q(0).$$

Thus

$$p(\infty)\mathfrak{s}\mathbf{1} = \mathbf{1} - \lim_{\theta \to 0} p(\theta)q(0)\mathbf{1} = \mathbf{1},$$

or

(11)
$$\sum_{i,j=1}^{N} p_i s_{ij} = 1,$$

an equation which it is impossible to satisfy unless some, and therefore all, of the p_i are positive.

Conversely, suppose that $p_j > 0$. Then, by Theorem 3.3, the canonical measure μ_{jj} of the p-function p_{jj} satisfies

(12)
$$\int x\mu_{jj}(dx) < \infty.$$

Now, for each j,

$$1 = \sum_k q_{jk}(\theta)\hat{p}_{kj}(\theta)$$

$$= \left\{\theta + \int (1 - e^{-\theta x})\mu_j(dx)\right\} \hat{p}_{jj}(\theta) - \sum_{k \neq j} \int e^{-\theta x}\lambda_{jk}(dx)\hat{p}_{kj}(\theta).$$

By Theorem 5.4, there exists a function $f_{kj} \geq 0$, integrable on $(0, \infty)$, such that $p_{kj} = f_{kj} * p_{jj}$, so that

$$\theta + \int (1 - e^{-\theta x})\mu_{jj}(dx) = \hat{p}_{jj}(\theta)^{-1}$$

$$= \theta + \int (1 - e^{-\theta x})\mu_j(dx)$$

$$- \sum_{k \neq j} \int e^{-\theta x}\lambda_{jk}(dx)f_{kj}(\theta).$$

Hence, on $(0, \infty)$,

(13) $$\mu_{jj}(dx) = \mu_j(dx) + \sum_{k \neq j} (f_{kj} * d\lambda_{jk})\, dx.$$

Multiplying by x and integrating over $(0, \infty)$, we have

$$\int x\mu_{jj}(dx) = \int x\mu_j(dx)$$
$$+ \sum_{k \neq j} \left\{ \int f_{kj}(x\,,\, dx \int x\lambda_{jk}(dx) + \lambda_{jk}[0,\,\infty) \int xf_{kj}(x)\, dx \right\}.$$

Since the left hand side is finite by (12), so is each term on the right hand side. Hence (5) holds for all i, and (6) holds for all pairs i, j with

$$\int_0^\infty f_{ji}(t)\, dt > 0.$$

This last condition is, however, always satisfied, since otherwise p_{ji} would be always zero and the assumption of irreducibility would be violated. ◆

The theorem then shows that, if m_i and l_{ij} are finite, then numbers π_j may be calculated using (7), which depend only on the numbers $\mu_i\{\infty\}$, $\lambda_{ij}[0, \infty)$, such that

$$p_j = c\pi_j$$

for some $c > 0$. The scalar multiple c may be determined by the normalising equation (11) in terms of the s_{ij}. Now

$$s_{ij} = l_{ij} \quad (i \neq j), \qquad s_{ii} = 1 + m_i,$$

so that (11) may be written

(14) $$\sum_{i=1}^{N} p_i \left(1 + m_i + \sum_{j \neq i} l_{ij} \right) = 1.$$

It is worth noting that the limits $p_{ij}(\infty)$ therefore depend on the measures μ_i, λ_{ij} of the canonical representation only through the numbers

$$\mu_i\{\infty\}, \qquad \lambda_{ij}[0, \infty), \qquad \int_{(0,\infty)} x\mu_i(dx), \qquad \int_{(0,\infty)} x\lambda_{ij}(dx),$$

This is very reminiscent of the basic ergodic theory of semi-Markov processes [68], and the resemblance is far from accidental; if all the μ_i are totally finite, the quasi-Markov chain Z is just a particular type of semi-Markov process.

A quite different approach to Theorem 5.6 may be found in [50].

5.5 NOTES

(i) The theory of quasi-Markov chains was originally developed under the name of 'linked systems of regenerative events', and most of the results of this chapter may be found in [39]. That paper also contains an application to the classification of theorems about Markov transition probabilities, the underlying structure of which later became clearer in the light of the developments to be described in the next chapter.

(ii) The analogue for a quasi-Markov chain of the additive process (4.1.9) has been studied by Neveu [61] under the name '*processus F*'.

(iii) A consequence of Theorem 5.2 is that the determinant of $\hat{p}(\theta)$ is strictly positive for all $\theta > 0$ (and has positive real part for all complex θ with Re $\theta > 0$). Thus in particular the resolvent matrix

$$(1) \qquad\qquad (\hat{p}_{ij}(\theta))$$

of a Markov chain always has the property that its principal minors are positive.

(iv) *Duality.* It is shown in [39] (see also [50]) that, if p is an irreducible standard p-matrix, there exist positive numbers m_j $(1 \leqslant j \leqslant N)$ such that, for all j,

$$(2) \qquad\qquad \sum_{i \neq j} m_i \lambda_{ij}[0, \infty) \leqslant m_j \mu_j\{\infty\}.$$

(Clearly $m_j = p_j$ is one such choice, so long as $p_j > 0$.) Then the matrix p^* defined by

$$(3) \qquad\qquad p_{ij}^*(t) = (m_j/m_i)p_{ji}(t)$$

is again a standard p-matrix.

As an application of this result, the reader may like to show that, in a quasi-Markov chain of order 2,

$$\int_0^\infty g_{12}(t)\, dt < \infty$$

if p_{21} is not identically zero.

(v) *Solidarity equivalence.* Call two distinct standard p-functions p_1 and p_2 S-equivalent if there exists an irreducible standard p-matrix with

$$p_{11}(t) = p_1(t), \qquad p_{22}(t) = p_2(t).$$

This defines a relation on $\mathscr{P}$ which has some affinities with the notion of E-equivalence considered in §2.7 (xv). Both relations imply a certain similarity in the behaviour of equivalent functions for large t. The relation of S-equivalence has been implicitly studied by Chung and others in proving 'solidarity theorems' for Markov chains.

(vi) It is important to note a difference in kind between the two cases of Theorem 5.4. To decide whether a function is a diagonal element of some standard p-matrix is to decide whether it belongs to $\mathscr{P}$, and this can in principle be determined by computing the canonical measure μ. But in the non-diagonal case it is necessary to decide whether there exists a representation of the form (3.11), and I know of no effective recipe for doing this. There is nothing canonical about the elements of this representation. The difficulty should not surprise us, since it has already arisen in the much simpler discrete time case (§1.6 (xv)).

CHAPTER 6

The Markov Characterisation Problem

6.1 THE PROBLEM AND ITS SOLUTION

This chapter describes the solution of the delicate problem of describing the class $\mathscr{PM}$, or equivalently, of establishing a criterion for a standard p-function to be a diagonal element of some Markov transition matrix. Since there is a one-to-one correspondence between standard p-functions and their canonical measures, the criterion may legitimately be expressed in terms of the canonical measure. It turns out that, roughly speaking, a standard p-function belongs to $\mathscr{PM}$ if and only if its canonical measure is reasonably smooth, and not too small.

Theorem 6.1. *Let p be a standard p-function with canonical measure μ. Then there exists a Markov chain with transition probabilities $(p_{ij}(t); i, j \in S)$ on a countable state space S, and a state $i \in S$ such that*

$$(1) \qquad\qquad p(t) = p_{ii}(t),$$

if and only if μ is absolutely continuous on $(0, \infty)$ and admits a lower semi-continuous density f which is either identically zero or satisfies, for some a, the inequalities

$$(2) \qquad f(t) > 0 \quad (0 < t < 1), \qquad f(t) \geqslant \mathrm{e}^{-at} \quad (t \geqslant 1).$$

The fact that absolute continuity of μ is necessary appears to have been known to Lévy [57], but the argument given here was shown to me by Reuter, and it yields without much difficulty the other necessary conditions (2). The sufficiency is proved by generalising a construction of Yuskevitch [76], using what Chung likes to call 'escalators'. The details take up the next three sections.

It should be noted that the atom (if any) of μ at ∞ plays no part in determining whether or not p belongs to $\mathscr{PM}$. A direct proof of this fact may be found in [43].

6.2 THE CANONICAL MEASURE OF A MARKOV p-FUNCTION

Throughout this section, $(p_{ij}(t); i, j \in S)$ will denote the transition probabilities of a standard Markov chain X on the countable state space S. For any pair of distinct states $i, j \in S$, the matrix

(1)
$$\begin{pmatrix} p_{ii}(t) & p_{ij}(t) \\ p_{ji}(t) & p_{jj}(t) \end{pmatrix}$$

is a standard p-matrix of order 2, and the results of §5.3 may therefore be invoked to write

(2)
$$p_{ij} = g_{ij} * p_{ii}, \qquad g_{ij} = {}_{i}p_{jj} * d\lambda_{ij},$$
$$p_{ji} = f_{ji} * p_{ii}, \qquad f_{ji} = {}_{i}p_{jj} * d\lambda_{ji}.$$

Here the expression ${}_{i}p_{jj}$ denotes the taboo probability defined (if the chain is separable) by

(3)
$$\quad {}_{i}p_{jj}(t) = \mathbf{P}_{j}\{X(s) \neq i \ (0 < s < t), X(t) = j\},$$

and the identification of this taboo probability is justified by (5.2.10). For $s, t > 0$,

$$p_{ii}(s + t) = \sum_{j} p_{ij}(s)p_{ji}(t)$$
$$= p_{ii}(s)p_{ii}(t) + \sum_{j \neq i} p_{ij}(s)p_{ji}(t),$$

and hence, for $\alpha, \beta > 0$, $\alpha \neq \beta$,

$$\frac{\hat{p}_{ii}(\alpha) - \hat{p}_{ii}(\beta)}{\beta - \alpha} = \hat{p}_{ii}(\alpha)\hat{p}_{ii}(\beta) + \sum_{j \neq i} \hat{p}_{ij}(\alpha)\hat{p}_{ji}(\beta)$$
$$= \hat{p}_{ii}(\alpha)\hat{p}_{ii}(\beta) + \sum_{j \neq i} \hat{g}_{ij}(\alpha)\hat{p}_{ii}(\alpha)\hat{f}_{ji}(\beta)\hat{p}_{ii}(\beta)$$

by (2), so that

$$(\beta - \alpha)^{-1}\{\hat{p}_{ii}(\beta)^{-1} - \hat{p}_{ii}(\alpha)^{-1}\} = 1 + \sum_{j \neq i} \hat{g}_{ij}(\alpha)\hat{f}_{ji}(\beta).$$

If μ is the canonical measure of p_{ii}, substitute

$$\hat{p}_{ii}(\theta)^{-1} = \theta + \int (1 - e^{-\theta x})\mu(dx)$$

to obtain

(4)
$$\int_{(0, \infty)} \frac{e^{-\alpha x} - e^{-\beta x}}{\beta - \alpha} \mu(dx) = \sum_{j \neq i} \hat{g}_{ij}(\alpha)\hat{f}_{ji}(\beta).$$

Letting $\beta \to \alpha$, and using the monotone convergence theorem, we have

$$\int x \, \mathrm{e}^{-\alpha x} \mu(dx) = \sum_{j \neq i} \hat{g}_{ij}(\alpha) \hat{f}_{ji}(\alpha)$$

$$= \int \mathrm{e}^{-\alpha x} \sum_{j \neq i} (g_{ij} * f_{ji}) \, dx.$$

Thus μ is absolutely continuous on $(0, \infty)$, and admits the density

(5) $$h(x) = x^{-1} \sum_{j \neq i} (g_{ij} * f_{ji})(x).$$

From (2), g_{ij} and f_{ji} are continuous except at the atoms of λ_{ij} and λ_{ji}. Thus their convolution is, by Fatou's lemma, lower semicontinuous, and hence $h(x)$ is lower semicontinuous on $(0, \infty)$.

Now suppose that h is not identically zero, and fix j so that

$$k = g_{ij} * f_{ji}$$

is not identically zero. From (2),

(6) $$k = \tilde{p} * \tilde{p} * d\lambda_{ij} * d\lambda_{ji},$$

where

$$\tilde{p} = {}_i p_{jj} \in \mathscr{P}.$$

Hence $\tilde{p}(t) > 0$ and, since $\tilde{p}$ is supermultiplicative (cf. §2.7 (xix))

$$\lim_{t \to \infty} t^{-1} \log \tilde{p}(t)$$

exists. Hence there exists b such that

$$\tilde{p}(t) \geqslant \tfrac{1}{2} \mathrm{e}^{-bt}$$

for all t, and so

$$(\tilde{p} * \tilde{p})(t) \geqslant \tfrac{1}{4} t \, \mathrm{e}^{-bt}.$$

It follows at once from (6) that, for some a,

$$t^{-1} k(t) > 0 \quad (0 < t < 1), \qquad t^{-1} k(t) \geqslant \mathrm{e}^{-at} \quad (t \geqslant 1),$$

and therefore that h satisfies (1.2), if it can be shown that, for all $\eta > 0$,

(7) $$\lambda_{ij}[0, \eta) > 0, \qquad \lambda_{ji}[0, \eta) > 0.$$

In order to prove (7), suppose to the contrary that, for some η,

$$\lambda_{ij}[0, \eta) = 0.$$

Then (2) implies that $p_{ij}(t) = 0$ for $t \leqslant \eta$, and hence, by the celebrated Lévy–Austin–Ornstein theorem [6], p_{ij} is identically zero. This however implies that $\lambda_{ij}[0, \infty) = 0$, which contradicts the assumption made in the choice of j.

Hence we have proved that the conditions of Theorem 6.1 are necessary for p to belong to $\mathscr{PM}$. For completeness, we include a proof of the Lévy–Austin–Ornstein theorem (which is essentially Chung's improvement on Ornstein's proof; an improvement of Chung's improvement of Austin's proof will be found in §7.3).

Theorem 6.2. *In a standard Markov chain, the function $p_{ij}(\cdot)$ is either identically zero, or is always positive on $(0, \infty)$.*

Proof. There is no loss of generality in taking the state space to be the positive integers, and $i = 1, j = 2$ (the diagonal case $i = j$ being trivial). Since p_{12} is continuous, and

$$p_{12}(t + u) \geqslant p_{12}(t)p_{22}(u),$$

there must exist $t_0 \in [0, \infty]$ such that

(8)
$$\begin{aligned} p_{12}(t) &= 0 \quad (0 < t \leqslant t_0) \\ p_{12}(t) &> 0 \quad (t_0 < t < \infty). \end{aligned}$$

The theorem asserts that t_0 is 0 or ∞; suppose on the contrary that $0 < t_0 < \infty$, so that

$$p_{12}(t) = 0 \quad (t \leqslant t_0), \qquad p_{12}(2t_0) = c > 0.$$

Since

$$\sum_{j=1}^{\infty} p_{ij}(t) = 1,$$

with all functions continuous in the compact interval $[0, t_0]$, the series is, by Dini's theorem, uniformly convergent, and there exists N such that

(9)
$$\sum_{j=N+1}^{\infty} p_{ij}(t) < \tfrac{1}{4}c \quad (0 \leqslant t \leqslant t_0).$$

Write $s = t_0/2N$ and, for $m \geqslant 1$,

$$A_m = \{k \geqslant 1; p_{1k}(ms) > 0\}.$$

Since

$$p_{1k}((m + 1)s) \geqslant p_{11}(s)p_{1k}(ms),$$

we have

$$A_m \subseteq A_{m+1};$$

write

$$B_1 = A_1, \qquad B_m = A_m - A_{m-1} \quad (m \geqslant 2).$$

If $k \notin A_m$, then

$$0 = p_{1k}(ms)$$
$$= \sum_{l=1}^{\infty} p_{1l}((m-1)s)p_{lk}(s)$$
$$= \sum_{l \in A_{m-1}} p_{1l}((m-1)s)p_{lk}(s),$$

so that

(10) $$p_{lk}(s) = 0 \qquad (l \in A_{m-1}, k \notin A_m)$$

Let $m \leqslant 2N$ and $k \notin A_m$. Then, for $n \geqslant 0$,

$$p_{1k}((n+1)s) = \left\{ \sum_{l \notin A_m} + \sum_{l \in B_m} + \sum_{l \in A_{m-1}} \right\} p_{1l}(ns)p_{lk}(s),$$

and the third sum vanishes by (10). Hence

$$\sum_{k \notin A_m} p_{1k}((n+1)s) = \left\{ \sum_{l \notin A_m} + \sum_{l \in B_m} \right\} p_{1l}(ns) \sum_{k \notin A_m} p_{lk}(s)$$
$$\leqslant \sum_{l \notin A_m} p_{1l}(ns) + \sum_{l \notin B_m} p_{1l}(ns),$$

or

$$\sum_{k \notin A_m} \{ p_{1k}((n+1)s) - p_{1k}(ns) \} \leqslant \sum_{l \in B_m} p_{1l}(ns).$$

Summing over n from 0 to $4N-1$, and noting that $1 \in A_m$,

$$\sum_{k \notin A_m} p_{1k}(4Ns) \leqslant \sum_{n=0}^{4N-1} \sum_{l \in B_m} p_{1l}(ns).$$

In particular, since $2 \notin A_m$ when $m \leqslant 2N$, and $p_{12}(4Ns) = c$, we have

(11) $$c \leqslant \sum_{n=0}^{4N-1} \sum_{l \in B_m} p_{1l}(ns).$$

The sets $B_1, B_2, \ldots, B_{2N}$ are disjoint, so that we can find N of them disjoint from $\{1, 2, \ldots, N\}$. If their union is B, then (11) implies that

$$Nc \leqslant \sum_{n=0}^{4N-1} \sum_{l \in B} p_{1l}(ns)$$
$$\leqslant \sum_{n=0}^{4N-1} \sum_{l=N+1}^{\infty} p_{1l}(ns)$$
$$< 4N \cdot \tfrac{1}{4}c$$

by (9). The contradiction completes the proof. ◆

6.3 THE YUSKEVITCH CONSTRUCTION AND ITS LIMITATIONS

In order to produce Markov transition functions with unpleasant analytic properties, Yuskevitch ([76], see also [31]) constructed Markov chains in the following way. The states are labelled

$$0,$$

(1)
$$(0, r) \ , \ r = 1, 2, \ldots,$$
$$(n, r) \ , n = 1, 2, \ldots, r = 1, 2, \ldots, k_n,$$

where the k_n are arbitrary positive integers. Positive constants $q, q_1, q_2, \ldots$ and a probability distribution $(b_n; n \geq 0)$ are chosen, and the progress of the chain, starting at 0, is as follows. It remains in 0 for an exponentially distributed time with mean q^{-1}, and then selects a non-negative integer n with probability b_n. It jumps to $(n, 1)$, and then proceeds through the states $(n, 1), (n, 2), \ldots$ in order, spending in each an exponentially distributed time with mean q_n^{-1}. On leaving (n, k_n), it returns to 0, and the whole process starts again (unless by chance $n = 0$, in which case it never returns to 0).

It is easy to see that this recipe does define a Markov chain (though a formal proof will not be given since the result will be superseded by a more general one proved in the next section). The canonical measure μ of the p-function p_{00} is easily computed; it has an atom $\mu\{\infty\} = qb_0$ and a density

(2)
$$q \sum_{n=1}^{\infty} b_n q_n^{k_n} t^{k_n - 1} e^{-q_n t} (k_n - 1)!$$

in $(0, \infty)$. By suitable choice of the k_n, b_n, q_n, it is possible to generate any canonical measure with a density of the form

(3)
$$f(t) = \sum_{m=0}^{\infty} \sum_{n=1}^{\infty} a_{mn} t^m e^{-nt},$$

so long as $a_{mn} \geq 0$ and

(4)
$$\sum_{m=0}^{\infty} \sum_{n=1}^{\infty} a_{mn} m! n^{-m-1} < \infty.$$

Indeed, the construction may be extended in a manner due to Kolmogorov in such a way that 0 may be an instantaneous state, in which case (4) may be weakened to the condition

(5)
$$\sum_{m=0}^{\infty} \sum_{n=1}^{\infty} a_{mn} m! [n^{-m-1} - (n + 1)^{-m-1}] < \infty.$$

This is equivalent to

$$\int_0^\infty (1 - e^{-t})f(t)\, dt < \infty,$$

which must of course be satisfied by the density of any canonical measure.

Thus the Yuskevitch construction is very general, and can be made to show that large classes of functions can be densities of canonical measures of functions in $\mathscr{PM}$. Exactly how general such functions may be is shown by the following theorem [47], which is a key step in the proof of Theorem 6.1.

Theorem 6.3. *In order that a function $f\colon (0, \infty) \to [0, \infty]$ be expressible in the form*

(6) $$f(t) = \sum_{m=0}^{\infty} \sum_{n=0}^{\infty} a_{mn} t^m\, e^{-nt},$$

where $a_{mn} \geqslant 0$, it is necessary and sufficient that it be lower semicontinuous, and either that it be identically zero, or that it satisfy the inequality

(7) $$f(t) \geqslant at^m\, e^{-nt}$$

for some $a > 0$, $m, n \geqslant 0$.

(Notice that (6) is allowed to include $n = 0$, since no requirement of integrability is imposed on f in this theorem. There is of course no question of the representation (6) being unique.)

Proof. The necessity of the conditions is obvious. To prove the sufficiency, let us (in this proof) describe a function as *good* if it can be expressed in the form (6) with $a_{mn} \geqslant 0$. Clearly the sum of a countable number of good functions is good, and the product of two good functions is good. It suffices then to prove that every lower semicontinuous function f with

$$\inf_{t>0} f(t) > 0$$

is good, for then every function satisfying (7) can be written

$$f(t) = [t^m\, e^{-nt}][f(t)/t^m\, e^{-nt}]$$

as the product of two good functions.

Suppose therefore that f is lower semicontinuous, and that

(8) $$f(t) \geqslant m > 0.$$

We may exclude the function $f(t) = \infty$ $(0 < t < \infty)$, which is trivially good (take $a_{mn} = 2^{n^2}$), and assume that $f(t) < \infty$ for at least one value of t. Then the functions ϕ_n defined on $(0, \infty)$ by

$$(9) \qquad \phi_n(t) = \inf_{s > 0} \{f(s) + n\, d(s, t)\},$$

where

$$(10) \qquad d(s, t) = |s^{\frac{1}{2}} - t^{\frac{1}{2}}|$$

is a metric on $(0, \infty)$, are finite for all t. Clearly

$$(11) \qquad m \leqslant \phi_1(t) \leqslant \phi_2(t) \leqslant \ldots \leqslant f(t),$$

and the lower semicontinuity of f implies that

$$(12) \qquad \lim_{n \to \infty} \phi_n(t) = f(t)$$

(cf. [60], Theorem 7.9). Moreover,

$$\phi_n(u) \leqslant f(s) + n\, d(s, u)$$
$$\leqslant \{f(s) + n\, d(s, t)\} + n\, d(u, t),$$

so that

$$\phi_n(u) \leqslant \phi_n(t) + n\, d(u, t),$$

and interchanging u and t we have

$$(13) \qquad |\phi_n(u) - \phi_n(t)| \leqslant n\, d(u, t);$$

ϕ_n is Lipschitz with respect to the metric d. Hence, if we set

$$f_1(t) = \phi_1(t) - \tfrac{1}{2}m,$$
$$f_n(t) = \phi_n(t) - \phi_{n-1}(t) + (\tfrac{1}{2})^n m \qquad (n \geqslant 2),$$

then (11), (12) and (13) show that the f_n are Lipschitz functions with

$$\inf_{t > 0} f_n(t) > 0,$$

and that

$$f = \sum_{n=1}^{\infty} f_n.$$

If therefore we can prove that f_n is good, the goodness of f will follow.

Accordingly it suffices to prove that every function f, satisfying

$$(14) \qquad \inf_{t > 0} f(t) > 0, \qquad |f(u) - f(t)| \leqslant c\, d(u, t)$$

for some c, is good. To do this, consider the function

(15) $$\psi_n(t) = \sum_{r=0}^{\infty} f(r/n)\pi_r(nt),$$

where as usual

$$\pi_r(\mu) = \mu^r e^{-\mu}/r!$$

and the convergence of (15) is guaranteed by (14). Then

$$
\begin{aligned}
|\psi_n(t) - f(t)| &= \left| \sum_{r=0}^{\infty} \{f(r/n) - f(t)\}\pi_r(nt) \right| \\
&\leqslant \sum_{r=0}^{\infty} |f(r/n) - f(t)|\pi_r(nt).
\end{aligned}
$$

But

$$
\begin{aligned}
|f(u) - f(t)| &\leqslant c|u^{\frac{1}{2}} - t^{\frac{1}{2}}| \\
&= c|u - t|/(u^{\frac{1}{2}} + t^{\frac{1}{2}}) \\
&\leqslant ct^{-\frac{1}{2}}|u - t|,
\end{aligned}
$$

and setting $u = r/n$

$$
\begin{aligned}
|\psi_n(t) - f(t)| &\leqslant ct^{-\frac{1}{2}} \sum_{r=0}^{\infty} \left| \frac{r}{n} - t \right| \pi_r(nt) \\
&\leqslant ct^{-\frac{1}{2}} \left[\sum_{r=0}^{\infty} \left(\frac{r}{n} - t \right)^2 \pi_r(nt) \right]^{\frac{1}{2}} \\
&= ct^{-\frac{1}{2}}(n^{-1}t)^{\frac{1}{2}} = cn^{-\frac{1}{2}},
\end{aligned}
$$

where we have used Schwarz's inequality and the formula for the variance of the Poisson distribution. Hence

$$\lim_{n \to \infty} \psi_n(t) = f(t)$$

uniformly in $t > 0$.

In view of (14) we can now choose n so that

(16) $$\tfrac{3}{4}f(t) \leqslant \psi_n(t) \leqslant \tfrac{5}{4}f(t),$$

and with this choice of n write

(17) $$g_1(t) = \tfrac{2}{3}\psi_n(t), \qquad f_1(t) = f(t) - g_1(t).$$

Then it is asserted that f_1 and g_1 enjoy the following five properties:

(i) $f = f_1 + g_1$,

(ii) $\inf f_1(t) > 0$,

(iii) $|f_1(u) - f_1(t)| \leqslant 3c\, d(u, t)$,

(iv) $f_1 \leqslant \tfrac{1}{2} f$,

(v) g_1 is good.

Of these, (i), (ii) and (iv) follow at once from (16) and (17), and (v) from the fact that

$$g_1(t) = \sum_{r=0}^{\infty} \left[\tfrac{2}{3} f\left(\frac{r}{n}\right) \frac{n^r}{r!} \right] t^r\, e^{-nt}.$$

To prove (iii), note that for $u < t$,

$$\psi_n(t) - \psi_n(u) = \sum_{r=0}^{\infty} f(r/n)[\pi_r(nt) - \pi_r(nu)]$$

$$= \sum_{r=0}^{\infty} f(r/n) \int_u^t n\pi_r'(nv)\, dv$$

$$= n \int_u^t \sum_{r=0}^{\infty} f(r/n)\{\pi_{r-1}(nv) - \pi_r(nv)\}\, dv$$

$$= n \int_u^t \sum_{r=0}^{\infty} \{f(r+1/n) - f(r/n)\}\pi_r(nv)\, dv.$$

From (14),

$$\left| f\left(\frac{r+1}{n}\right) - f\left(\frac{r}{n}\right) \right| \leqslant c \left| \left(\frac{r+1}{n}\right)^{\frac{1}{2}} - \left(\frac{r}{n}\right)^{\frac{1}{2}} \right|$$

$$= cn^{-\frac{1}{2}}\{(r+1)^{\frac{1}{2}} + r^{\frac{1}{2}}\}^{-1}$$

$$\leqslant cn^{-\frac{1}{2}}(r+1)^{-\frac{1}{2}},$$

so that

$$|\psi_n(t) - \psi_n(u)| \leqslant cn^{\frac{1}{2}} \int_u^t \sum_{r=0}^{\infty} (r+1)^{-\frac{1}{2}}\pi_r(nv)\, dv.$$

But

$$\sum_{r=0}^{\infty} (r+1)^{-\frac{1}{2}}\pi_r(nv) \leqslant \left[\sum_{r=0}^{\infty} (r+1)^{-1}\pi_r(nv) \right]^{\frac{1}{2}}$$

$$= \left[\sum_{r=0}^{\infty} (nv)^{-1}\pi_{r+1}(nv) \right]^{\frac{1}{2}}$$

$$\leqslant (nv)^{-\frac{1}{2}},$$

so that

$$|\psi_n(t) - \psi_n(u)| \leqslant c \int_u^t v^{-\frac{1}{2}} \, dv$$
$$= 2c \, d(u, t).$$

Hence, for all t, u,

$$|f_1(u) - f_1(t)| \leqslant |f(u) - f(t)| + \tfrac{2}{3}|\psi_n(u) - \psi_n(t)|$$
$$\leqslant c \, d(u, t) + \tfrac{2}{3} \cdot 2c \, d(u, t)$$
$$\leqslant 3c \, d(u, t).$$

Notice in particular that (ii) and (iii) show that f_1 has the same properties (14) as were assumed of f. Thus we may repeat the process to give a decomposition $f_1 = f_2 + g_2$, and so on by induction, to give sequences $f = f_0, f_1, f_2, \ldots$ and $g_1, g_2, \ldots$ such that

   (i) $f_{n-1} = f_n + g_n$,

   (ii) $\inf f_n(t) > 0$,

   (iii) $|f_n(u) - f_n(t)| \leqslant 3^n c \, d(u, t)$,

   (iv) $f_n \leqslant \tfrac{1}{2} f_{n-1}$,

   (v) g_n is good.

From (i) and (iv),

$$f = \sum_{n=1}^{\infty} g_n$$

is a sum of good functions, and is therefore good. ◆

6.4 ESCALATORS

The discussion of the last section indicates that the p-function with canonical measure μ will belong to $\mathscr{PM}$ if μ admits on $(0, \infty)$ a lower semicontinuous density satisfying (3.7). This condition is stronger than (1.2), which has been proved in §6.2 to be necessary, and will now be proved sufficient, for p to belong to $\mathscr{PM}$. In fact, since a lower semicontinuous function attains its lower bound on every compact set, a function satisfying (1.2) will also satisfy (3.7) if and only if, for some m,

(1) $$f(t) \geqslant t^m$$

for small t. It is known that there are functions in $\mathscr{PM}$ whose canonical measures violate (1), the simplest example being a divergent pure birth process with instantaneous return [46].

The fact that the result of the Yuskevitch construction satisfies (1) is a reflection of the finite number of states through which the process passes between visits to the distinguished state. The key to removing the condition is therefore to replace Yuskevitch's finite strings of states by infinite strings which the process negotiates in a finite time. These have been variously described as 'flashes' or 'escalators', and have for some years played a major role in the construction of pathological Markov chains.

Suppose that a process moves in order through an infinite succession of states $s_1, s_2, s_3, \ldots$, remaining in s_j for an exponentially distributed time T_j with mean q_j^{-1}. Then the total time to traverse this string is

$$T = T_1 + T_2 + \ldots,$$

and, for $\theta > 0$,

$$\mathbf{E}(e^{-\theta T}) = \prod_{j=1}^{\infty} \mathbf{E}(e^{-\theta T_j})$$

$$= \prod_{j=1}^{\infty} \left(\frac{q_j}{q_j + \theta} \right).$$

Thus T will be almost certainly finite if and only if this infinite product converges, which will happen if and only if

(2)
$$\sum_{j=1}^{\infty} q_j^{-1} < \infty.$$

Under this condition, let G be the distribution function of

$$T' = T_2 + T_3 + \ldots.$$

Then, for $t > 0$,

$$\mathbf{P}(T \leqslant t) = \mathbf{P}(T_1 + T' \leqslant t)$$
$$= \mathbf{E}\{G(t - T_1)\}$$
$$= \int_0^t G(t - u) q_1 e^{-q_1 u} \, du.$$

Hence T has probability density

$$f(t) = \int_0^t q_1 e^{-q_1(t-s)} \, dG(s)$$
$$= q_1 e^{-q_1 t} \int_0^t e^{q_1 s} \, dG(s).$$

Thus f is continuous, and there exists a number $t_0 \geqslant 0$ (the greatest solution of $G(t_0) = 0$) such that

$$f(t) = 0 \qquad (t \leqslant t_0)$$
$$> 0 \qquad (t > t_0).$$

Suppose (to obtain a contradiction) that $t_0 > 0$, so that $\mathbf{P}(T > t_0) = 1$. Then, for $\theta > 0$,

$$e^{-\theta t_0} \geqslant \mathbf{E}(e^{-\theta T}) = \prod_{j=1}^{\infty} (1 + q_j^{-1}\theta)^{-1}.$$

In view of (2), we can choose J so that

$$\sum_{j=J+1}^{\infty} q_j^{-1} \leqslant \tfrac{1}{2} t_0,$$

and then

$$\prod_{j=J+1}^{\infty} (1 + q_j^{-1}\theta) \leqslant \exp \sum_{j=J+1}^{\infty} q_j^{-1}\theta \leqslant \exp (\tfrac{1}{2}\theta t_0),$$

so that

$$e^{-\theta t_0} \geqslant \prod_{j=1}^{J} (1 + q_j^{-1}\theta)^{-1} e^{-\frac{1}{2}\theta t_0},$$

or

$$e^{\frac{1}{2}\theta t_0} \leqslant \prod_{j=1}^{J} (1 + q_j^{-1}\theta).$$

This inequality holds for all θ, and implies that

$$e^{\frac{1}{2}\theta t_0} = O(\theta^J)$$

for large θ, which is impossible if $t_0 > 0$.

Symbolically, f may be thought of as an infinite convolution

(3) $$f = \phi_1 * \phi_2 * \ldots,$$

where

$$\phi_j(t) = q_j e^{-q_j t},$$

and (2) is satisfied. We shall describe any function f of the form

(4) $$f = a\phi_1 * \phi_2 * \ldots,$$

where $a > 0$, as an *escalator function*. Then every escalator function is positive and continuous in $(0, \infty)$ (and indeed by repeating the argument can be seen to have continuous derivatives of all orders*). Our key result is that any function satisfying the conditions of Theorem 6.1 is a countable sum of escalator functions.

Theorem 6.4. *Let* $f: (0, \infty) \to [0, \infty]$ *be lower semicontinuous and strictly positive, and suppose that* $f(t) \geqslant e^{-at}$ $(t \geqslant 1)$. *Then there exists a countable collection* $\{f_\alpha; \alpha \in A\}$ *of escalator functions such that*

$$(5) \qquad\qquad f = \sum_A f_\alpha.$$

For the proof we need two lemmas.

Lemma 1. *Let* f *be an escalator function, and* g *a lower semicontinuous function satisfying* (3.7). *Then there exists a countable collection* $\{h_\beta; \beta \in B\}$ *of escalator functions such that*

$$(6) \qquad\qquad fg = \sum_B h_\beta.$$

Proof. In view of Theorem 6.3, we need only prove the lemma in the special case $g(t) = t^m e^{-nt}$. Moreover, this special case will follow by induction on m and n if we can establish the two very special cases $g(t) = t$ and $g(t) = e^{-t}$. The reader will easily check, using Laplace transforms, that when $g(t) = t$,

$$fg = \sum_{\beta=1}^{\infty} (aq_\beta)^{-1} \phi_\beta * \phi_1 * \phi_2 * \ldots,$$

and that, when $g(t) = e^{-t}$,

$$fg = \bar{a}\bar{\phi}_1 * \bar{\phi}_2 * \ldots,$$

where

$$\bar{a} = a \prod_{j=1}^{\infty} (1 + q_j^{-1}), \qquad \bar{\phi}_j(t) = (1 + q_j) e^{-(1+q_j)t}.$$

Hence the lemma is true in these special cases, and is therefore true in general. ◆

Lemma 2. *Let* ω *be any positive non-decreasing function on* $(0, \infty)$. *Then there exists an escalator function* f *with*

$$(7) \qquad\qquad f(t) \leqslant \omega(t)$$

for all $t > 0$.

---

* An extended account of the theory of escalator functions, from a purely analytic point of view, may be found in [74].

Proof. We shall choose the constants $q_1, q_2, \ldots$ so as to make

$$f = \phi_1 * \phi_2 * \ldots$$

satisfy (7). If

$$g = \phi_2 * \phi_3 * \ldots,$$

then

$$f(t) = \int_0^t q_1 \, e^{-q_1(t-s)} g(s) \, ds$$

$$\leqslant q_1 \int_0^t g(s) \, ds \leqslant q_1.$$

Hence, if we choose

$$q_1 = \min \{\omega(\tfrac{1}{2}), 1\},$$

then (7) holds trivially when $t \geqslant \tfrac{1}{2}$. In what follows we therefore assume that $t < \tfrac{1}{2}$, and denote by β the unique integer with

(8) $$2^{-\beta-1} < t < 2^{-\beta}.$$

The remaining q_j will be chosen in terms of a non-decreasing sequence $\nu_1, \nu_2, \ldots$ of positive integers, where ν_α is the smallest integer ν with

(9) $$\nu \geqslant -11 \log \omega(2^{-\alpha-1}).$$

The q_j are to be defined by the formula

$$q_j = 2^\alpha \nu_\alpha \quad \left(2 + \sum_{i=1}^{\alpha-1} \nu_i \leqslant j \leqslant 1 + \sum_{i=1}^{\alpha} \nu_i\right),$$

and (2) is satisfied since

$$\sum_{j=2}^{\infty} q_j = \sum_{\alpha=1}^{\infty} (2^\alpha \nu_\alpha)^{-1} \nu_\alpha = \sum_{\alpha=1}^{\infty} 2^{-\alpha} < \infty.$$

For all $\theta > 0$,

$$f(t) = \int_0^t q_1 \, e^{-q_1(t-s)} g(s) \, ds$$

$$\leqslant q_1 \int_0^t g(s) \, ds$$

$$\leqslant q_1 \, e^{\theta t} \int_0^\infty g(s) \, e^{-\theta s} \, ds$$

$$= q_1 \, e^{\theta t} \prod_{j=2}^{\infty} (1 + q_j^{-1}\theta)^{-1}$$

$$= q_1 \, e^{\theta t} \prod_{\alpha=1}^{\infty} (1 + 2^{-\alpha}\nu_\alpha^{-1}\theta)^{-\nu_\alpha}.$$

Since $q_1 \leqslant 1$,

$$\log f(t) \leqslant \theta t - \sum_{\alpha=1}^{\infty} v_\alpha \log (1 + 2^{-\alpha} v_\alpha^{-1} \theta)$$

$$\leqslant \theta - \sum_{\alpha=\beta}^{\infty} v_\alpha \log (1 + 2^{-\alpha} v_\alpha^{-1} \theta).$$

Now

$$\log (1 + x) \geqslant x - \tfrac{1}{2}x^2, \qquad (x \geqslant 0),$$

so that

$$\log f(t) \leqslant \theta t - \sum_{\alpha=\beta}^{\infty} v_\alpha (2^{-\alpha} v_\alpha^{-1} \theta - 2^{-2\alpha-1} v_\alpha^{-2} \theta^2)$$

$$= \theta t - 2^{-\beta+1} \theta + \theta^2 \sum_{\alpha=\beta}^{\infty} 2^{-2\alpha-1} v_\alpha^{-1}$$

$$\leqslant \theta t - 2\theta t + \theta^2 v_\beta^{-1} \sum_{\alpha=\beta}^{\infty} 2^{-2\alpha-1}$$

$$= -\theta t + \tfrac{1}{3}\theta^2 v_\beta^{-1} 2^{-2\beta+1}$$

$$\leqslant -\theta t + \tfrac{8}{3}\theta^2 t^2 v_\beta^{-1},$$

using (8). This is valid for all $\theta > 0$, and in particular for

$$\theta = 3v_\beta/16t,$$

whence

$$\log f(t) \leqslant -\tfrac{3}{32} v_\beta$$

$$\leqslant \log \omega(2^{-\beta-1})$$

$$\leqslant \log \omega(t),$$

and (7) follows. $\blacklozenge$

Proof of Theorem 6.4. If f satisfies the conditions of the theorem, the function

$$\omega(t) = \inf_{t \leqslant u \leqslant 1} f(u)$$

is positive and non-decreasing in (0, 1]. Define $\omega(t) = \omega(1)$ for $t > 1$, and find an escalator function h with

$$h(t) \leqslant \omega(t) \leqslant f(t)$$

on (0, 1]. Then

$$g(t) = f(t)/h(t)$$

is lower semicontinuous and satisfies (3.7) (with $m = 0$). Lemma 1 then shows that $f = gh$ is a countable sum of escalator functions. $\blacklozenge$

Armed with these results, we can now complete the proof of Theorem 6.1. The case $f \equiv 0$ being trivial, we have to show that the p-function whose canonical measure μ is given by

(10)
$$\mu(dx) = f(x)\, dx \qquad (0 < x < \infty)$$
$$\mu\{\infty\} = a$$

belongs to $\mathscr{PM}$ whenever $a \geqslant 0$ and f is lower semicontinuous and satisfies (1.2). By Theorem 6.4, f is a countable sum of escalator functions, and we only require the following result to complete the argument.

Theorem 6.5. *Let a be a non-negative number, and $\{f_\alpha; \alpha \in A\}$ a countable collection of escalator functions such that*

(11)
$$\sum_A \int_0^\infty (1 - e^{-z}) f_\alpha(x)\, dx < \infty.$$

Then there is a Markov chain and a state 0 such that the canonical measure of the p-function p_{00} is given by (10), where

(12)
$$f = \sum_A f_\alpha.$$

Proof. The escalator function f_α may be written in the form

$$f_\alpha = a_\alpha \phi_{1\alpha} * \phi_{2\alpha} * \ldots,$$

where

$$\phi_{j\alpha}(t) = q_{j\alpha}\, e^{-q_{j\alpha}t}.$$

It is convenient to write

$$\Phi_{rs\alpha} = \phi_{r\alpha} * \phi_{r+1,\alpha} * \phi_{r+2,\alpha} * \ldots * \phi_{s\alpha},$$

so that

$$f_\alpha = a_\alpha \Phi_{1\infty\alpha}.$$

Add a further index δ to A, and write

$$a_\delta = a, \qquad q_{j\delta} = 1, \qquad B = A \cup \{\delta\}.$$

Notice that there is no escalator function f_δ, since

$$\sum_{j=1}^\infty q_{j\delta}^{1-} = \infty.$$

Because of (11), there is a standard p-function p with canonical measure given by (10) and (12); we have to prove that $p \in \mathscr{PM}$. This is done by modifying the construction described in the last section by replacing each finite string of states by an escalator, to give a 'bouquet of escalators', in the following way.

The state space S consists of the distinguished state 0, together with the ordered pairs

$$(\alpha, r), \qquad \alpha \in B, r = 1, 2, 3, \ldots .$$

Functions p_{ij} $(i, j \in S)$ are defined by the equations

$$p_{00} = p$$

$$p_{0,(\beta,s)} = a_\beta q_{s\beta}^{-1} p * \Phi_{1s\beta} \qquad (\beta \in B)$$

$$p_{(\alpha,r),0} = \Phi_{r\infty\alpha} * p \qquad (\alpha \in A)$$

(13)
$$p_{(\alpha,r),(\beta,s)} = a_\beta q_{s\beta}^{-1} \Phi_{r\infty\alpha} * p * \Phi_{1s\beta}$$

$$+ \langle q_{s\beta}^{-1} \Phi_{rs\beta} \rangle \qquad (\alpha \in A, \beta \in B)$$

$$p_{(\delta,r),0} = 0$$

$$p_{(\delta,r),(\beta,s)} = 0 \qquad (\beta \in A)$$

$$p_{(\delta,r),(\delta,s)} = \langle q_{s\delta}^{-1} \Phi_{rs\delta} \rangle,$$

where the terms in angular brackets are to be read as zero unless $r \leqslant s$ and (in the first case) $\alpha = \beta$. It is then a matter of straightforward computation to show that the continuous functions p_{ij} satisfy

$$p_{ij}(t) \geqslant 0, \qquad p_{ij}(0) = \delta_{ij},$$

(14)
$$\sum_{j \in S} p_{ij}(t) = 1,$$

$$p_{ij}(t + u) = \sum_{k \in S} p_{ik}(t) p_{kj}(u).$$

The computations are best accomplished using Laplace transforms; we illustrate them by the case $i = j = 0$ of the last equation.

For $\theta, \psi > 0$, $\theta \neq \psi$, we have

$$\sum_{k \in S} \hat{p}_{0k}(\theta) \hat{p}_{k0}(\psi) = \hat{p}_{00}(\theta) \hat{p}_{00}(\psi) + \sum_{\alpha \in A} \sum_{r=1}^{\infty} \hat{p}_{0,(\alpha,r)}(\theta) \hat{p}_{(\alpha,r),0}(\psi)$$

$$= \hat{p}(\theta) \hat{p}(\psi) + \sum_A \sum_{r=1}^{\infty} a_\alpha q_{r\alpha}^{-1} \hat{p}(\theta) \prod_{m=1}^{r} \hat{\phi}_{m\alpha}(\theta) \prod_{n=r}^{\infty} \hat{\phi}_{n\alpha}(\psi) \hat{p}(\psi).$$

This expression includes a term

$$\hat{\phi}_{r\alpha}(\theta) \hat{\phi}_{r\alpha}(\psi) = (\theta - \psi)^{-1} q_{r\alpha} \{ \hat{\phi}_{r\alpha}(\psi) - \hat{\phi}_{r\alpha}(\theta) \},$$

so that

$$\sum_{k \in S} \hat{p}_{0k}(\theta)\hat{p}_{k0}(\psi) = \hat{p}(\theta)\hat{p}(\psi)\left\{1 + (\theta - \psi)^{-1}\sum_A a_\alpha \sum_{r=1}^{\infty}\left[\prod_{m=1}^{r-1}\hat{\phi}_{m\alpha}(\theta)\prod_{n=r}^{\infty}\hat{\phi}_{n\alpha}(\psi)\right.\right.$$
$$\left.\left. - \prod_{m=1}^{r}\hat{\phi}_{m\alpha}(\theta)\prod_{n=r+1}^{\infty}\hat{\phi}_{n\alpha}(\psi)\right]\right\}$$
$$= \hat{p}(\theta)\hat{p}(\psi)\left\{1 + (\theta - \psi)^{-1}\sum_A a_\alpha\left[\prod_{n=1}^{\infty}\hat{\phi}_{n\alpha}(\psi) - \prod_{m=1}^{\infty}\hat{\phi}_{m\alpha}(\theta)\right]\right\}$$
$$= \hat{p}(\theta)\hat{p}(\psi)\left\{1 + (\theta - \psi)^{-1}\sum_A [\hat{f}_\alpha(\psi) - \hat{f}_\alpha(\theta)]\right\}$$
$$= \hat{p}(\theta)\hat{p}(\psi)\left\{\theta - \psi + \sum_A \int (e^{-\psi x} - e^{-\theta x})f_\alpha(x)dx\right\}(\theta - \psi)^{-1}$$
$$= \hat{p}(\theta)\hat{p}(\psi)\left\{\theta - \psi + \int (e^{-\psi x} - e^{-\theta x})\mu(dx)\right\}(\theta - \psi)^{-1}$$
$$= \hat{p}(\theta)\hat{p}(\psi)\{\hat{p}(\theta)^{-1} - \hat{p}(\psi)^{-1}\}(\theta - \psi)^{-1}$$
$$= \int_0^{\infty} p(s)\frac{e^{-\theta s} - e^{-\psi s}}{\psi - \theta}\,ds$$
$$= \int_0^{\infty}\int_0^{\infty} p(t + u)\,e^{-\theta t - \psi u}\,dt\,du.$$

Hence, inverting the double Laplace transforms, and invoking the continuity of the functions (13), we have

$$p_{00}(t + u) = \sum_{k \in S} p_{0k}(t)p_{k0}(u).$$

The other cases of (14) are exactly similar, and show that (13) defines the transition probabilities of a standard Markov chain. In particular, $p = p_{00}$ belongs to $\mathscr{PM}$. ◆

With this result, the last link of the chain of proof of Theorem 6.1 is forged, and the proof is complete.

6.5 THE INFINITELY DIVISIBLE CASE

Theorem 6.1 provides a complete solution to the problem of characterising $\mathscr{PM}$, and enables one to decide, by examining its canonical measure, whether a given standard p-function can arise from a Markov chain. There are however functions which can be shown to belong to $\mathscr{P}$ otherwise than by calculating the canonical measure, and for which such calculation may not be feasible. For these it may be a non-trivial problem to test for membership of $\mathscr{PM}$.

The most notable class for which this difficulty arises is that studied in §2.5, where it was shown that, if ϕ is non-negative, continuous and concave on $[0, \infty)$, with $\phi(0) = 0$, then

$$(1) \qquad p(t) = e^{-\phi(t)}$$

defines a standard p-function. Such functions are exactly those which can be written in the form

$$(2) \qquad p(t) = \exp\left\{-\int_{(0,\infty]} \min\,(t, x)\nu(dx)\right\},$$

where the measure ν satisfies

$$(3) \qquad \int_{(0,\infty]} \min\,(1, x)\nu(dx) < \infty.$$

It is natural to ask under what conditions the function (2) belongs to $\mathscr{PM}$. Necessary and sufficient conditions are not yet known, but the following theorem gives sufficient conditions which cannot be too far from being necessary.

Theorem 6.6. *The function* (2) *belongs to* $\mathscr{PM}$ *if the measure ν is absolutely continuous on* $(0, \infty)$, *with lower semicontinuous density which is positive in* $(0, A)$ *for some* $A > 0$.

Proof. Suppose first that ν is supported by the finite interval $(0, \Delta)$, that $0 < \lambda = \nu(0, \Delta) < \infty$, and that the density $\lambda k(x)$ is continuous and strictly positive on $(0, \Delta)$. Then

$$(4) \qquad p(t) = \exp\left\{-\lambda \int_0^\Delta \min\,(t, x)k(x)\,dx\right\}.$$

In §2.5 a regenerative phenomenon with p-function (4) was constructed by means of a Poisson process $0 < X_1 < X_2 < \ldots$ of rate λ and a sequence of independent variables Y_n with probability density $k(y)$; the phenomenon occurs on the complement of the random set

$$(5) \qquad \mathscr{G} = \bigcup_{n=1}^{\infty} (X_n, X_n + Y_n).$$

From (4) the phenomenon is stable (with $q = \lambda$) and recurrent, so that the canonical measure takes the form

$$(6) \qquad \mu(dx) = \lambda\,dG(x),$$

where G is the distribution function of the intervals of non-occurrence, i.e. of the lengths of the connected components of $\mathscr{G}$.

Label these components, in order, as $\mathcal{G}_1, \mathcal{G}_2, \ldots$. Then $\mathcal{G}_1$ is of the form

$$\mathcal{G}_1 = (X_1, X_N + Y_N)$$

for some $N \geqslant 1$, and the left hand endpoint of $\mathcal{G}_2$ is X_M for some $M > N$. Thus we may write

(7)
$$G = \sum_{1 \leqslant n < m} G_{nm},$$

where

(8)
$$G_{nm}(x) = \mathbf{P}(L \leqslant x, N = n, M = m)$$

and

$$L = X_N - X_1 + Y_N.$$

Since the variables $X_n' = X_n - X_1 \; (n \geqslant 2)$ form a Poisson process, we may therefore write

$$G_{nm}(x + h) - G_{nm}(x)$$

$$= \int \cdots \int \lambda^{m-1} e^{-\lambda x_m} k(y_1) k(y_2) \cdots$$

$$\times k(y_{m-1}) \, dx_2 \ldots dx_m \, dy_1 \ldots dy_{m-1} + o(h),$$

where the integral is taken over the set delineated by the inequalities

$$0 < x_1 < x_2 < \ldots < x_m,$$
$$x_r + y_r < x \quad (r = 1, 2, \ldots, n-1, n+1, \ldots, m-1),$$
$$x < x_n + y_n < x + h < x_m,$$

$(x_1 = 0)$, and by the condition that the set

$$\bigcup_{r=1}^{m-1} (x_r, x_r + y_r)$$

should be connected. Letting $h \to 0$ shows that G_{nm} is absolutely continuous, with density

(9)
$$g_{nm}(x) = \int \cdots \int \lambda^{m-1} e^{-\lambda x_m} k(y_1) \cdots$$

$$\times k(y_{n-1}) k(x - x_n) k(y_{n+1}) \cdots k(y_{m-1})$$

$$\times dx_2 \, dx_3 \ldots dx_m \, dy_1 \ldots dy_{n-1} \, dy_{n+1} \ldots dy_{m+1},$$

which is easily seen to be everywhere continuous, and to be positive in $(0, n\Delta)$. Hence from (7) G has a lower semicontinuous density which is everywhere positive.

In order to prove that $p \in \mathscr{PM}$, it therefore suffices by Theorem 6.1 to show that, for some α,

$$(10) \qquad\qquad g(x) \geqslant e^{-\alpha x}$$

for large x. To do this, write $\delta = \frac{1}{5}\Delta$, and evaluate (9), with $n = m - 1$, over the smaller region of integration delineated by the inequalities

$$(2r - 1)\delta < x_r < 2r\delta \qquad (2 \leqslant r \leqslant m - 1),$$
$$3\delta < y_r < 4\delta \qquad (1 \leqslant r \leqslant m - 2),$$
$$x < x_m,$$

to give

$$g(x) \geqslant g_{m-1,m}(x)$$
$$\geqslant \lambda^{m-2} e^{-\lambda x}(k_0\delta)^{m-2}\delta^{m-3} \int_{(2m-3)\delta}^{(2m-2)\delta} k(x - x_{m-1})\, dx_{m-1},$$

where

$$k_0 = \inf \{k(y); 3\delta \leqslant y \leqslant 4\delta\}$$

and $x \geqslant 2m\delta$. Hence for

$$2m\delta \leqslant x \leqslant (2m + 1)\delta$$

we have

$$g(x) \geqslant A\eta^m,$$

where A is a constant and

$$\eta = \lambda k_0 \delta^2 e^{-\lambda\delta}.$$

Thus (10) holds for

$$x \in \bigcup_{m=1}^{\infty} [2m\delta, (2m + 1)\delta].$$

A similar argument deals with the intervals $[(2m + 1)\delta, (2m + 2)\delta]$, and completes the proof that (4) defines an element of $\mathscr{PM}$, under the special assumptions made at the beginning of the proof.

More generally, let ν be any measure satisfying the conditions of the theorem. Then (by the argument used at the beginning of the proof of Theorem 6.3) ν may be expressed as a countable sum

$$\nu = \sum_{n=0}^{\infty} \nu_n,$$

where ν_0 is supported by $\{\infty\}$ and, for $n \geqslant 1$, ν_n is either zero or is totally finite, supported by a finite interval $(0, \Delta_n)$, and has a positive continuous density on $(0, \Delta_n)$. Then we have already shown that, for $n \geqslant 1$,

$$p^{(n)}(t) = \exp\left\{-\int \min(t, x)\,\nu_n(dx)\right\}$$

is in $\mathscr{PM}$, and the result is trivially true when $n = 0$. Hence $p(t)$ may be expressed by the convergent infinite product

(11) $$p(t) = \prod_{n=0}^{\infty} p^{(n)}(t)$$

of functions in $\mathscr{PM}$.

Now construct, for each n, a standard Markov chain $(p_{ij}^{(n)}(t))$ on the non-negative integers so that

$$p_{00}^{(n)}(t) = p^{(n)}(t).$$

Let S denote the countable set of infinite sequences

$$i = (i_0, i_1, \ldots)$$

of non-negative integers such that $i_k = 0$ for all but a finite number of k, and define functions p_{ij} $(i, j \in S)$ by

(12) $$p_{ij}(t) = \prod_{n=1}^{\infty} p_{i_n j_n}^{(n)}(t).$$

It is then a routine matter (cf. Chung's account of Blackwell's example [6]) to check that (12) defines a family of standard Markov transition functions on S. Clearly the p-function of the state $(0, 0, \ldots)$ is given by (11), so that p belongs to $\mathscr{PM}$. ◆

This theorem provides the richest known source of explicit examples of members of $\mathscr{PM}$, and some of these are rather surprising. For example, if ν is supported by the finite interval $(0, \Delta)$, then $p(t)$ is constant in $t \geqslant \Delta$;

$$p(t) = \exp\left\{-\int x\nu(dx)\right\} \qquad (t \geqslant \Delta).$$

6.6 NOTES

(i) It might not at first sight be clear that Theorem 6.1 is effective, in the sense of providing an algorithm whereby a function in $\mathscr{P}$ may be tested for membership of $\mathscr{PM}$. This is because, although μ is uniquely determined by p, its density is not, and a given measure will often admit many different lower semicontinuous densities.

The answer to this objection is the following recipe. For the canonical measure μ, let

(1) $$f(t) = \sup \phi(t),$$

where the supremum is taken over all continuous functions ϕ satisfying

$$\int_a^b \phi(t)\, dt \leqslant \mu(a, b)$$

for all $a < b$. Then f is lower semicontinuous. If f is a density for μ, then it is the maximal lower semicontinuous density for μ, and thus satisfies (1.2) if any such density does. On the other hand, if f is not a density for μ, then μ admits no lower semicontinuous density. In other words, to test for membership of $\mathscr{PM}$ we may compute f, check that it is a density for μ, and then verify that it satisfies (1.2). (For the arguments supporting these assertions, see [47].)

(ii) The mode of proof of Theorem 6.1 yields further information. It shows for example that, if p belongs to $\mathscr{PM}$, there exists a chain in which all states, except perhaps 0, are stable, with $p = p_{00}$. The construction may be modified [47] to show that the chain may be chosen so that all states, except perhaps 0, are instantaneous, or indeed to achieve an arbitrary mixture of stable and instantaneous states. Thus a knowledge of the function p_{00} tells us nothing about the other states in the chain.

(iii) We might describe the construction of §6.4 as being one of a system of escalators arranged in parallel. I conjectured in [47] that it might be possible to replace this by a single escalator in series with a parallel system of finite strings of states. This was foolish; such a chain would have the untypical property that the density of the canonical measure were infinitely differentiable.

(iv) The trick used in the proof of Theorem 6.3 is a modification of that used by Bernstein in his celebrated proof of the Weierstrass approximation theorem. He used the binomial distribution where we have used the Poisson distribution. Unfortunately, in our case the uniformity of convergence is both essential and more difficult to establish; the introduction of the metric d is effectively equivalent to the use of the variance-stabilising transformation of the Poisson distribution.

(v) *The reversible case.* Theorem 6.1 having solved the (diagonal) characterisation problem for general Markov chains, one may pose similar problems for restricted classes of chains. For example, the argument at the end of §2.3 leads to the conclusion that the functions p_{ii} in q-bounded chains (those with bounded infinitesimal generator) are exactly the functions of the class $\mathscr{Q}$. Another example is to be found in [44], where it is shown that the functions p_{ii} in reversible chains are exactly the completely monotonic functions

$$(2) \qquad\qquad p(t) = \int e^{-tx} \gamma(dx),$$

where γ is a probability measure on $[0, \infty)$.

(vi) *The non-diagonal problem.* It is an obvious problem to prove a companion theorem to Theorem 6.1, characterising the non-diagonal functions p_{ij} in standard Markov chains, although in the light of the remarks made in §§1.6 (xv), 5.6 (vi) this cannot be expected to be in any sense effective. It was in fact shown in [47] that a function q is of the form p_{ij} for some pair i, j of distinct states in some Markov chain if and only if q can be written in the form

$$(3) \qquad\qquad q = \acute{p} * \, d\lambda * \grave{p},$$

where $\acute{p}$ and $\grave{p}$ belong to $\mathscr{PM}$, and the totally finite measure λ satisfies

$$\int_0^\infty \acute{p}(t) \, dt \; \lambda[0, \infty) \leqslant 1,$$

and has a lower semicontinuous density on $(0, \infty)$ which is identically zero or satisfies (1.2).

Although there is no obvious way of deciding whether a given function q admits a representation (3) with the given properties, the theorem can be used to establish properties of the non-diagonal transition probabilities for a Markov chain. For instance, it is an interesting analytic exercise to use (3) to prove Ornstein's theorem that q has a continuous derivative in $[0, \infty)$.

(vii) *Quasi-Markov characterisations.* The most systematic way of arriving at the results described in (vi) is to ask the question; *given a collection of functions p_{ij} $(i, j = 1, 2, \ldots, N)$, under what conditions can this be extended to a collection p_{ij} $(i, j = 1, 2, \ldots)$ which are the transition probabilities of a standard Markov chain?* Theorem 6.1 solves the problem in case $N = 1$, and the methods used to prove it may readily be extended to cover the general case. The result is that necessary and sufficient conditions are that $(p_{ij}; i, j = 1, 2, \ldots, N)$ should be a standard p-matrix, and that, in the canonical representation of Theorem 5.2, the measures μ_i and λ_{ij} should all admit lower semicontinuous densities on $(0, \infty)$ which are either zero or satisfy (1.2).

6

Once this has been proved, the result of (vi) follows at once from Theorem 5.4.

(viii) Theorem 6.6 gives a sufficient condition for an infinitely divisible p-function p to belong to $\mathscr{PM}$. A necessary condition is that p be continuously differentiable, which is the same as saying that ν has no atoms in $(0, \infty)$.

Another necessary condition follows from noting that, for any function in $\mathscr{PM}$, either

$$(4) \qquad\qquad p(s + t) > p(s)p(t)$$

for all $s, t > 0$, or $p(t) = e^{-\alpha t}$ for some $\alpha \geqslant 0$. This is because

$$p_{00}(s + t) - p_{00}(s)p_{00}(t) = \sum_{j \neq 0} p_{0j}(s)p_{j0}(t),$$

where the functions on the right hand side are either always positive or identically zero. Applying this to the infinitely divisible case shows that either $\nu(0, \infty) = 0$ or $\nu(0, \eta) > 0$ for all $\eta > 0$.

Thus the gap between the necessary and the sufficient conditions is not wide, and should be possible to close by strengthening the analytic arguments.

A complete solution to this problem would have consequences for the arithmetic of the semigroup $\mathscr{PM}$. For example, it would probably imply that an element of $\mathscr{PM}$ which is infinitely divisible in $\mathscr{P}$ is also infinitely divisible in $\mathscr{PM}$, a result which, if true, would be non-trivial since (cf. [12], [13]) $\mathscr{PM}$ is not hereditary in $\mathscr{P}$ (elements of $\mathscr{PM}$ admit factors in $\mathscr{P} - \mathscr{PM}$).

A key step in improving Theorem 6.6 would be a proof that an infinitely divisible function $p \in \mathscr{PM}$ has a representation (5.2) where ν admits a lower semicontinuous density, which will be true if p' can be shown to be the indefinite integral of a lower semicontinuous function. An unsuccessful attempt to prove this result went like this. Let f be the lower semicontinuous density of the canonical measure of p. Then by (4.3.10),

$$(5) \qquad p(s + t) = p(s)p(t) + \int_0^s \int_0^t p(s - u)p(t - v)f(u + v)\, du\, dv.$$

Differentiating formally with respect to s and t in turn, we obtain

$$p''(s + t) = p'(s)p'(t) + f(s + t)$$
$$+ \int_0^t p'(t - v)f(s + v)\, dv + \int_0^s p'(s - u)f(u + t)\, du$$
$$+ \int_0^s \int_0^t p'(s - u)p'(t - v)f(u + v)\, du\, dv.$$

Writing $s = T - t$, and integrating with respect to t on $(0, T)$, we obtain

$$Tp''(T) = (p' * p')(T) + F(T) + 2(p' * F)(T) + (p' * p' * F)(T),$$

where

$$(6) \qquad\qquad F(t) = tf(t).$$

The derivation of this identity is purely formal, and cannot in general be justified. However, it should be noted that both p' and F are integrable on $(0, T)$, so that the right hand side always makes sense for almost all t, and a rigorous Laplace transform argument shows that the identity is valid in the integrated sense

$$(7) \quad \int_0^T t \, dp'(t) = \int_0^T \{(p' * p') + F + 2(p' * F) + (p' * p' * F)\} \, dt.$$

If it could be shown that the integrand of the right hand side were lower semicontinuous, our immediate goal would be achieved. Now, since $p' \leqslant 0$, the first, second and fourth terms are lower semicontinuous, but the third is upper semicontinuous. The argument therefore breaks down.

It does however throw up the interesting identity (7), which has other uses. It is valid for all p-functions, if f is replaced by the canonical measure μ. More precisely, if the measure κ on $(0, \infty)$ is defined by

$$(8) \qquad\qquad \kappa(dx) = x\mu(dx),$$

then

$$(9) \quad \int_0^T t \, dp'(t)$$
$$= \int_0^T \{(p' * p') + 2(p' * d\kappa) + (p' * p' * d\kappa)\} \, dt + \kappa(0, T],$$

where p' is the right derivative of p.

An application of (9) is to the problem mentioned in §3.6 (vii). If $p(\infty) > 0$, then κ is totally finite, and p is of bounded variation V on $(0, \infty)$. Hence (9) shows that

$$\int_0^\infty |t \, dp'(t)| \leqslant V^2 + (1 + V)^2\kappa(0, \infty) < \infty,$$

so that

$$\int_1^\infty |dp'(t)| < \infty.$$

This shows that

$$\lim_{t \to \infty} p'(t)$$

exists, and since p is bounded the limit is zero. Hence

$$\lim_{t \to \infty} p'(t) = 0$$

for all standard p-functions with $p(\infty) > 0$.

CHAPTER 7

Markov Processes on General State Spaces

7.1 INTRODUCTION

While it is undeniably true that the most significant application of the theory of regenerative phenomena to Markov theory is to the situation in which there are only a countable number of possible states, the theory does also have relevance to more general Markov processes. In this chapter some of the features of this relevance are outlined.

We follow fairly closely the analysis of [45], except in one important respect. In that paper the state space was assumed to be a perfectly general measurable space (S, Σ), and this generality gave rise to two difficulties which complicated the argument and (though to a surprisingly small extent) weakened the conclusions. The first is that it is not always possible to assert that, to every transition function on (S, Σ), there corresponds a Markov process. The other is that regular conditional distributions may not always exist.

Both difficulties may be avoided by assuming a fairly pleasant topo-logical structure for (S, Σ), and to do so will make the arguments more transparent without unreasonably restricting their generality. We shall therefore assume without further comment that *S is a polish space* (a separable topological space admitting a complete metric) *and that Σ is the σ-algebra of Borel subsets of S.* We will usually write S instead of (S, Σ).

As usual, a *Markov transition function* on S means a function $P_t(x, E)$ defined for $t > 0$, $x \in S$, $E \in \Sigma$ and such that

(i) for fixed t, x, $P_t(x, \cdot)$ is a probability measure on S,
(ii) for fixed t, E, $P_t(\cdot, E)$ is a measurable function on S,
(iii) for all s, t, x, E,

(1)
$$P_{s+t}(x, E) = \int_S P_s(x, dy) P_t(y, E).$$

The problem is to examine the properties of $P_t(x, E)$ as a function of the 'time' variable t.

For any probability measure π on S, there is a Markov process $(X(t)$; $t \geqslant 0)$ with initial distribution π and transition function $P_t(x, E)$, its finite-dimensional distributions given by the formula

(2) $\mathbf{P}\{X(t_r) \in A_r (0 \leqslant r \leqslant n)\}$

$$= \int_{A_0} \cdots \int_{A_n} \pi(dx_0) \prod_{r=1}^{n} P_{t_r - t_{r-1}}(x_{r-1}, dx_r),$$

for $0 = t_0 < t_1 < \ldots < t_n$. The initial measure π will usually be taken as ϵ_x for some $x \in S$, when $\mathbf{P}$ will be written $\mathbf{P}_x$.

Let x be any pont in S. Then the function $\phi \colon S \to \{0, 1\}$ defined by

(3) $$\phi(x) = 1, \quad \phi(y) = 0 \quad (y \neq x),$$

is measurable, so that

(4) $$Z(t) = \phi\{X(t)\}$$

is a stochastic process taking the values 0 and 1. From (2) we have

$$\mathbf{P}_x\{Z(t_r) = 1 \ (1 \leqslant r \leqslant n)\} = \mathbf{P}_x\{X(t_r) \in \{x\} \ (1 \leqslant r \leqslant n)\}$$

$$= \prod_{r=1}^{n} P_{t_r - t_{r-1}}(x, \{x\}).$$

Thus Z is a regenerative phenomenon, under $\mathbf{P}_x$, with p-function

(5) $$p(t) = P_t(x, \{x\}).$$

In many Markov processes the p-function (5) will be identically zero, and the theory of p-functions has then nothing to say (cf. §4.6 (xv)). But there are processes in which, for some or even all of the states x, the p-function is standard, and then the theory is very relevant. With this in mind we make the following definition.

Definition. A point $x \in S$ is a *standard state* if

(6) $$\lim_{t \to 0} P_t(x, \{x\}) = 1.$$

Thus, if x is standard, its p-function (5) belongs to $\mathscr{P}$. The converse also applies, in the sense that any function in $\mathscr{P}$ can arise in this way. Indeed, if $p \in \mathscr{P}$, the Markov process F constructed in §4.3, and taking values in the polish space $[0, \infty]$, has 0 as a standard state with p-function p. Hence we have the following characterisation theorem.

Theorem 7.1. *In order that a function p should be the p-function of a standard state in some Markov process, it is necessary and sufficient that p belong to $\mathscr{P}$.*

It should be noted that, in this theorem, nothing is said about states other than x. In particular, the process F has no other standard states. If all states are required to be standard, the problem is quite different, and will be discussed in §§7.3, 7.4. Before tackling this problem, we consider what can be said about the function $P_t(x, E)$, where x is standard and E a general subset of S.

7.2 THE LAST EXIT DECOMPOSITION

In this section it will be assumed that x is a fixed standard state with p-function p; no assumption is made about the other states. We first need a lemma noted by Kendall [28].

Lemma. *For any $E \in \Sigma$,*

$$(1) \qquad |P_t(x, E) - P_s(x, E)| \leqslant 1 - p(|s - t|),$$

so that $P_t(x, E)$ is a uniformly continuous function of t.
Proof. Suppose without loss of generality that $s = t + u > t$. Then by (1.1)

$$
\begin{aligned}
P_s(x, E) &= \int_S P_u(x, dy) P_t(y, E) \\
&\geqslant P_u(x, \{x\}) P_t(x, E) \\
&= p(u) P_t(x, E).
\end{aligned}
$$

Moreover,

$$
\begin{aligned}
P_s(x, E) - p(u) P_t(y, E) &= \int_{S-\{x\}} P_u(x, dy) P_t(y, E) \\
&\leqslant \int_{S-\{x\}} P_u(x, dy) \\
&= 1 - p(u),
\end{aligned}
$$

so that

$$
\begin{aligned}
-\{1 - p(u)\}\{1 - P_t(x, E)\} &\leqslant P_s(x, E) - P_t(x, E) \\
&\leqslant \{1 - p(u)\} P_t(x, E)
\end{aligned}
$$

and thus

$$|P_s(x, E) - P_t(x, E)| \leqslant 1 - p(u). \qquad \blacklozenge$$

Theorem 7.2. *There exists a non-negative function $g(t, E)$ defined for $t > 0$, $E \in \Sigma$, such that*

(i) *for each $E \in \Sigma$, $g(\cdot, E)$ is Lebesgue measurable,*
(ii) *for each $t > 0$, $g(t, \cdot)$ is a totally finite Borel measure on S, with $g(t, \{x\}) = 0$ and*

(2) $$g(t, S) = \mu(t, \infty],$$

where μ is the canonical measure of p, and
(iii) *for $t > 0$, $x \in S$, $E \in \Sigma$,*

(3) $$P_t(x, E) = p(t)\epsilon_x(E) + \int_0^t p(t - s)g(s, E)\, ds.$$

The *last exit decomposition* (3) is of course well known in the countable case [6].

Proof. Since

$$P_t(x, E) = p(t)\epsilon_x(E) + P_t(x, E - \{x\}),$$

it will suffice to consider only the case when E does not contain x. For any $h > 0$, and any integer $n \geqslant 1$,

$$P_{nh}(x, E) = \int_S \ldots \int_S P_h(x, dy_1)P_h(y_1, dy_2) \ldots P_h(y_{n-1}, E).$$

The range of integration is the union of the disjoint sets

$$A_m = \{(y_1, y_2, \ldots, y_{n-1}); y_{n-m} = x, y_j \neq x\ (n - m < j < n)\}$$
$$(m = 1, 2, \ldots, n - 1)$$

and

$$A_n = \{(y_1, y_2, \ldots, y_{n-1}); y_j \neq x\ (1 \leqslant j < n)\},$$

so that

$$P_{nh}(x, E) = \sum_{m=1}^{n} p[(n - m)h]Q_h(m, E),$$

where

$$Q_h(m, E) = \int_{S'} \ldots \int_{S'} P_h(x, dz_1)P_h(z_1, dz_2) \ldots P_h(z_{m-1}, E)$$

and $S' = S - \{x\}$. Hence

(4) $$P_{nh}(x, E) = \int_{[0, nh]} p(nh - s)G_h(ds, E),$$

where $G_h(\cdot, E)$ is the measure on $[0, \infty)$ which assigns mass $Q_h(m, E)$ to the point mh ($m = 1, 2, \ldots$).

Notice that, for any Borel set $B \subseteq [0, \infty)$, $G_h(B, \cdot)$ is a measure on S', and in particular that

(5) $$G_h(B, E) \leqslant G_h(B, S').$$

In the special case $E = S'$, (4) reduces to

$$1 - p(nh) = \int p(nh - s)G_h(ds, S').$$

Multiplying by $e^{-\theta nh}(\theta > 0)$ and summing over n, we have

$$\int e^{-\theta t}G_h(dt, S') = \left\{(1 - e^{-\theta h})\sum_{n=0}^{\infty} p(nh)\, e^{-\theta nh}\right\}^{-1} - 1,$$

so that

$$\lim_{h \to 0} \int e^{-\theta t}G_h(dt, S') = [\theta \hat{p}(\theta)]^{-1} - 1$$

$$= \theta^{-1} \int (1 - e^{-\theta x})\mu(dx)$$

$$= \hat{m}(\theta),$$

where as usual $m(t) = \mu(t, \infty]$. Since this holds for all $\theta > 0$, $G_h(\cdot, S')$ converges as $h \to 0$, weakly on every finite interval, to the measure $G(\cdot, S')$ with density m.

Now fix $E \in \Sigma$. Because of (5) we can choose a sequence of values of h, tending to zero, along which $G_h(\cdot, E)$ converges, weakly on finite intervals, to a measure $G(\cdot, E)$, and

$$G(\cdot, E) \leqslant G(\cdot, S').$$

Since $G(\cdot, S')$ is absolutely continuous with respect to Lebesgue measure, so is $G(\cdot, E)$, so that there is a function $g(\cdot, E)$ such that

$$g(t, E) \leqslant g(t, S') = m(t)$$

and

$$G(B, E) = \int_B g(t, E)\, dt.$$

In (4) let $h \to 0$ through the selected sequence, and take $n = [t/h]$, using the inequality (1), to give

$$P_t(x, E) = \int p(t - s)G(ds, E)$$

$$= \int_0^t p(t - s)g(s, E)\, ds,$$

proving (3).

Now suppose that

$$E = \bigcup_1^\infty E_n$$

is a dissection of E into disjoint Borel sets E_n. Then

$$P_t(x, E) = \sum_{n=1}^\infty P_t(x, E_n),$$

so that

$$\int_0^t p(t - s)g(s, E)\, ds = \sum_{n=1}^\infty \int_0^t p(t - s)g(s, E_n)\, ds.$$

Taking Laplace transforms,

$$\hat{p}(\theta)\hat{g}(\theta, E) = \sum_{n=1}^\infty \hat{p}(\theta)\hat{g}(\theta, E_n),$$

which shows that

(6) $$g(t, E) = g(t, E_n)$$

for almost all t. Since S is a polish space, we can therefore choose a function $\tilde{g}(t, E)$ such that $\tilde{g}(t, \cdot)$ is a measure on S' for all $t > 0$, and such that

$$g(t, E) = \tilde{g}(t, E)$$

for all t outside a null set depending on E. Then

$$P_t(x, E) = \int_0^t p(t - s)\tilde{g}(s, E)\, ds$$

for $E \in \Sigma, E \subseteq S'$. To complete the proof, define

$$\tilde{g}(t, E) = \tilde{g}(t, E \cap S')$$

for all $E \in \Sigma$, and drop the bar on $\tilde{g}$.   ◆

A consequence of Theorem 7.2 is a characterisation of the functions $P_.(x, E)$, of which Theorem 7.1 is a special case.

Theorem 7.3. *Let F be a function on* $(0, \infty)$. *In order that there should exist a Markov process on some polish space S, a standard state* $x \in S$, *and a Borel subset* $E \subseteq S$, *such that*

$$F(t) = P_t(x, E), \tag{7}$$

it is necessary and sufficient that F be expressible in the form

$$F(t) = \epsilon p(t) + \int_0^t p(t - s)g(s) \, ds, \tag{8}$$

where $\epsilon = 0$ *or* 1, p *belongs to* $\mathscr{P}$, *and g is a measurable function with*

$$0 \leqslant g(t) \leqslant \mu(t, \infty], \tag{9}$$

where μ *is the canonical measure of p.*

Proof. The necessity follows from Theorem 7.2; notice that p is the p-function of x, and that $\epsilon = 0$ or 1 according as $x \notin E$ or $x \in E$.

To prove the sufficiency, suppose first that μ has no atoms, so that the function

$$\psi(t) = \inf \{x > 0; m(x) \leqslant m(t) - g(t)\}$$

is measurable and satisfies

$$\psi(t) \geqslant t, \qquad m[\psi(t)] = m(t) - g(t).$$

Consider the Markov process K defined in §4.3, and define E by

$$
\begin{aligned}
E &= \{(b, f); b > 0, 0 < f < \psi(b) - b\} &&\text{if } \epsilon = 0, \\
E &= \{(b, f); b > 0, 0 < f < \psi(b) - b\} \cup \{(0, 0)\} &&\text{if } \epsilon = 1.
\end{aligned}
$$

Then $x = (0, 0)$ is a standard state with p-function p and, by (4.3.12),

$$
\begin{aligned}
P_t(x, E) &= \iint_E p(t - b)\Gamma(db \, df) \\
&= \epsilon p(t) + \iint_{\substack{b > 0 \\ 0 < f < \psi(b) - b}} p(t - b)\Gamma_0(db \, df) \\
&= \epsilon p(t) + \int_0^t p(t - b)\{m(b) - m[\psi(b)]\} \, db \\
&= \epsilon p(t) + \int_0^t p(t - b)g(b) \, db \\
&= F(t).
\end{aligned}
$$

When μ has atoms, this construction must be modified by considering the process (K, L), where L is a random function taking independent constant values on the different excursions from $(0, 0)$; we omit the details (cf. [45]).   ◆

Like the other non-diagonal characterisations, Theorem 7.3 has a disturbingly non-effective character. It does however have a number of significant consequences, some of which will be explored in §7.5.

7.3 PURELY DISCONTINUOUS PROCESSES

According to a rather unfortunate terminology, the Markov process with transition function $P_t(x, E)$ is said to be *purely discontinuous* if every state is standard, i.e. if

(1) $$\lim_{t \to 0} P_t(x, \{x\}) = 1$$

for every $x \in S$. (A better term, in accordance with Markov chain usage, would be 'standard', but this adjective has acquired a quite different connotation in general Markov theory.)

The processes constructed to prove Theorems 7.1 and 7.3 are not purely discontinuous, and it is therefore possible that the assumption that all states are standard restricts the transition function more severely than do those two theorems. This is indeed true, and moreover all the special properties of the countable case extend to the more general purely discontinuous situation. Our main result is Theorem 7.5, but before proving this we establish the general case of the Lévy–Austin–Ornstein theorem.

Theorem 7.4. *Let $P_t(x, E)$ be the transition function of a purely discontinuous Markov process on a polish space S, and let $x \in S$, $E \in \Sigma$. Then either $P_t(x, E) > 0$ for all $t > 0$, or $P_t(x, E) = 0$ for all $t > 0$.*

Proof. Since

$$P_{s+t}(x, E) \geqslant P_s(x, \{x\})P_t(x, E),$$

the positivity of $P_t(x, E)$ for any t assures it for all larger values of t. Since $P.$ is continuous, there exists a uniquely defined number $u(x, E) \in [0, \infty]$ such that

$$P_t(x, E) = 0 \quad (0 < t \leqslant u(x, E)),$$
$$P_t(x, E) > 0 \quad (u(x, E) < t).$$

The theorem asserts that $u(x, E)$ is either 0 or ∞. To prove this, suppose on the contrary that, for some x, E,

$$(2) \qquad\qquad 0 < u(x, E) < \infty.$$

We shall hold x and E fixed, and write

$$u(y) = u(y, E),$$
$$u_0 = u(x),$$
$$v(y) = \min [u(y), u_0].$$

Clearly u and v are measurable. Let $X(t)$ be a Markov chain with the given transition function, and $X(0) = x$, and write

$$Y_0(t) = v\{X(t)\}.$$

Then $Y_0(0) = u_0$, and for $0 \leqslant t < t + h$,

$$\mathbf{P}\{Y_0(t + h) = Y_0(t)\}$$
$$\geqslant \int_S P_t(x, dy) P_h(y, \{y\}) \to 1 \qquad (h \to 0)$$

by the dominated convergence theorem applied to (1). Hence Y_0 is continuous in probability, and so possesses a separable and measurable version Y, also satisfying

$$0 \leqslant Y(t) \leqslant u_0 = Y(0).$$

If $0 < h < v(y)$, then since $v(y) \leqslant u(y)$,

$$0 = P_{v(y)}(y, E) = \int_S P_h(y, dz) P_{v(y)-h}(z, E).$$

But

$$P_{v(y)-h}(z, E) > 0$$

whenever $v(z) < v(y) - h$, so that

$$P_h(y, \{z; v(z) < v(y) - h\}) = 0,$$

and this also holds trivially when $h \geqslant v(y)$. Hence

$$\mathbf{P}\{Y(t + h) < Y(t) - h\} = \int_S P_t(x, dy) P_h(y, \{z; v(z) < v(y) - h\})$$
$$= 0.$$

Since Y is separable, it follows that (outside a fixed null event which may be ignored),

$$Y(t + h) \geqslant Y(t) - h$$

for all $t, h \geqslant 0$, so that the process

$$Z(t) = Y(t) + t$$

is non-decreasing.

If

$$\Delta(t, h) = \mathbf{P}\{Y(t + h) \neq Y(t)\},$$

we have seen that

$$\lim_{h \to 0} \Delta(t, h) = 0$$

for all t so that, by dominated convergence,

$$\lim_{h \to 0} \int_0^\infty \Delta(t, h)\, e^{-t}\, dt = 0.$$

Thus we may choose a sequence (h_n), tending to zero, such that

$$\sum_n \int_0^\infty \Delta(t, h_n)\, e^{-t}\, dt < \infty,$$

and Fubini's theorem implies that, for all t outside a null set N,

$$\sum_n \Delta(t, h_n) < \infty.$$

The Borel–Cantelli lemma shows that, for each $t \notin N$,

$$\mathbf{P}\{Y(t + h_n) = Y(t) \text{ for all sufficiently large } n\} = 1.$$

Hence the measurable process

$$W(t) = \limsup_{n \to \infty} h_n^{-1}\{Z(t + h_n) - Z(t)\}$$

satisfies

$$\mathbf{P}\{W(t) = 1\} = 1$$

for all $t \notin N$. Fubini's theorem shows that all realisations of W outside an event Λ of probability zero have

$$W(t) = 1 \text{ for almost all } t.$$

Since Z is non-decreasing, its derivative Z' exists for almost all t, and where it exists, $W = Z'$. Hence, outside Λ,

$$Z'(t) = 1 \text{ for almost all } t.$$

Since Z is non-decreasing,

$$Z(t) - Z(0) \geqslant \int_0^t Z'(s)\,ds = t$$

outside Λ, so that

$$Y(t) \geqslant u_0.$$

Thus we have proved that

$$P_t(x, \{y; u(y) < u_0\}) = 0.$$

In particular, since u vanishes on E,

$$P_t(x, E) = 0$$

for all t, which contradicts (2). ◆

The problem of characterising the functions $P_t(x, E)$ in a purely discontinuous process is therefore a non-trivial extension of that solved in Theorem 7.3. We shall concentrate on the diagonal case $E = \{x\}$, for which the problem has a striking and easily formulated solution.

7.4 THE DIAGONAL CASE

According to Theorem 7.1, any standard p-function can be associated with a standard state in some Markov process. The situation is radically different if all states are required to be standard; the only p-functions which can then arise are those in $\mathscr{PM}$.

Theorem 7.5. *Let $P_t(x, E)$ be the transition function of a purely discontinuous Markov process on a polish space S. Then the p-function*

$$p(t) = P_t(x, \{x\})$$

of any state x belongs to $\mathscr{PM}$.

Proof. Fix $x \in S$ and write $S' = S - \{x\}$ and p for the p-function of x. For any $y \in S'$ the process $\psi\{X(t)\}$, where $\psi(x) = 1$, $\psi(y) = 2$, $\psi(z) = 0$ ($z \neq x, y$) is a quasi-Markov chain with standard p-matrix

$$\begin{pmatrix} P_t(x, \{x\}) & P_t(x, \{y\}) \\ P_t(y, \{x\}) & P_t(y, \{y\}) \end{pmatrix}$$

Hence by Theorem 5.3, there exists a non-negative measurable function $f(\cdot, y)$ on $(0, \infty)$ such that

$$(1) \qquad P_t(y, \{x\}) = \int_0^t f(u, y)p(t - u)\,du.$$

Moreover, f can be written

$$(2) \qquad f(u, y) = \int_{[0,u]} p(u - v, y)\lambda(dv, y),$$

where $p(\cdot, y) \in \mathscr{P}$ and $\lambda(\cdot, y)$ is a totally finite Borel measure on $[0, \infty)$. In particular, $f(\cdot, y)$ is right-continuous. Hence

$$f(t, y) = \lim_{n \to \infty} n\{F(t + n^{-1}, y) - F(t, y)\},$$

where, using (1) and (3.6.16),

$$F(t, y) = \int_0^t f(u, y)\,dy$$
$$= P_t(y, \{x\}) + \int_0^t P_u(y, \{x\})m(t - u)\,du.$$

It follows that f is jointly measurable in its two arguments.

Now consider the function

$$(3) \qquad w(s, t) = \int_{S'} g(s, dy)f(t, y),$$

where g is defined by Theorem 7.2. Using the lemma of Feller quoted in [28] we have, for $\alpha, \beta > 0$, $\alpha \neq \beta$,

$$\omega(\alpha, \beta) = \int_0^\infty \int_0^\infty w(s, t)\,e^{-\alpha s - \beta t}\,ds\,dt$$
$$= \int_{S'} \hat{g}(\alpha, dy)\hat{f}(\beta, y).$$

From (2.3),

$$\hat{g}(\alpha, E) = \hat{p}(\alpha)^{-1}\hat{P}_\alpha(x, E),$$

and from (1),

$$\hat{f}(\beta, y) = \hat{p}(\beta)^{-1}\hat{P}_\beta(y, \{x\})$$

so that

$$\omega(\alpha, \beta) = [\hat{p}(\alpha)\hat{p}(\beta)]^{-1} \int_{S'} \hat{P}_\alpha(x, dy)\hat{P}_\beta(y, \{x\})$$

$$= [\hat{p}(\alpha)\hat{p}(\beta)]^{-1} \int_0^\infty \int_0^\infty e^{-\alpha s - \beta t} \int_{S'} P_s(x, dy)P_t(y, \{x\})\, ds\, dt$$

$$= [\hat{p}(\alpha)\hat{p}(\beta)]^{-1} \int_0^\infty \int_0^\infty e^{-\alpha s - \beta t}[p(s + t) - p(s)p(t)]\, ds\, dt$$

$$= [\hat{p}(\alpha)\hat{p}(\beta)]^{-1} \left\{ \frac{\hat{p}(\alpha) - \hat{p}(\beta)}{\beta - \alpha} - \hat{p}(\alpha)\hat{p}(\beta) \right\}$$

$$= \int \frac{e^{-\alpha x} - e^{-\beta x}}{\beta - \alpha}\, \mu(dx)$$

$$= \int \int e^{-\alpha s - \beta t} \Gamma_0(ds\, dt),$$

in the notation of Chapter 4. Hence Γ_0 has a density $w(s, t)$, which means that μ must have a density h (say) on $(0, \infty)$, and

$$w(s, t) = h(s + t)$$

for almost all (s, t). Thus

(4) $$h(s + t) = \int_{S'} g(s, dy)f(t, y)$$

for almost all (s, t).

Let

$$C = \{y \in S'; P_1(y, \{x\}) > 0\}.$$

Then Theorem 7.4 shows that, for any $t > 0$,

$$P_t(y, \{x\}) > 0 \qquad \text{if } y \in C,$$
$$P_t(y, \{x\}) = 0 \qquad \text{if } y \notin C.$$

Hence, from (1), $f(u, y) = 0$ if $y \notin C$. On the other hand, if $y \in C$, f is not identically zero, and (2) shows that, for some t_0,

$$f(u, y) = 0 \qquad (u \leqslant t_0)$$
$$f(u, y) > 0 \qquad (u > t_0).$$

If $t_0 > 0$, $P_t(y, \{x\}) = 0$ in $t \leqslant t_0$, which is impossible. Thus $f(u, y) > 0$ for all $u > 0$, $y \in C$.

7

If h is almost everywhere zero, p belongs trivially to $\mathcal{PM}$. For the rest of the proof we therefore assume that h is positive on a set of positive measure. Since by (4)

$$(5) \qquad th(t) = \int_C \int_0^t g(s, dy) f(t - s, y) \, ds$$

for almost all t, this implies that

$$(6) \qquad \int_0^t g(s, C) \, ds > 0$$

on a t-set of positive measure. If there is a positive t_0 with

$$\int_0^{t_0} g(s, C) \, ds = 0,$$

then $P_{t_0}(x, C) = 0$, and a contradiction results. Thus (5) and (6) may be taken to hold for all $t > 0$. The argument now proceeds exactly as in §6.2 to show that h is lower semicontinuous, and that

$$h(t) \geqslant e^{-\alpha t}$$

for some α and large t. Hence Theorem 6.1 shows that p belongs to $\mathcal{PM}$. ◆

7.5 NOTES

(i) According to Theorem 7.2,

$$(1) \qquad \gamma(t, \cdot) = g(t, \cdot)/m(t)$$

is, for each $t > 0$, a probability measure on S'. It is to be interpreted as the distribution of the state of the process after a time t has elapsed since the last exit from x. (See [6] for the proof for a countable state space.)

(ii) The effect of Theorem 7.3 is to associate, with each p in $\mathcal{P}$, a convex cone $\mathcal{T}_p$ consisting of those functions F which can be expressed in the form

$$(2) \qquad F(t) = \int_0^t p(t - s) g(s) \, ds,$$

where

$$(3) \qquad 0 \leqslant g(t) \leqslant \mu(t, \infty].$$

The possible functions $P_{\cdot}(x, E)$ with $x \notin E$ are then just the members of

$$\mathscr{T} = \bigcup_{p \in \mathscr{P}} \mathscr{T}_p.$$

It is instructive to examine $\mathscr{T}_p$ in a particular case. Let

$$p(t) = e^{-\lambda t}$$

$(\lambda > 0)$, so that

$$\mu = \lambda \epsilon_\infty.$$

Then (2) becomes

$$F(t) = e^{-\lambda t} \int_0^t e^{\lambda s} g(s)\, ds,$$

with

$$0 \leqslant g(t) \leqslant \lambda.$$

Then $\mathscr{T}_p$ consists of all functions F which are absolutely continuous and satisfy

(4) $$0 \leqslant F' + \lambda F \leqslant \lambda.$$

It follows that any absolutely continuous function F with

(5) $$\operatorname{ess\,sup} \left(-\frac{F'}{F}\right) < \infty, \qquad \operatorname{ess\,sup} \left(\frac{F'}{1-F}\right) < \infty$$

belongs to $\mathscr{T}$.

(iii) *Analytic properties*. The example just cited shows that it is very difficult to find analytic properties enjoyed by all transition functions $P_{\cdot}(x, E)$. For instance, there are classical results [28] giving sufficient conditions for the existence of the finite limit

$$\lim_{t \to 0} t^{-1} P_t(x, E).$$

This is the same as asserting the existence and finiteness of

$$\lim_{t \to 0} t^{-1} \int_0^t g(s)\, ds,$$

which is certainly not implied by (3).

At the other end of the scale, however, the existence of

$$p(\infty) = \lim_{t \to \infty} p(t)$$

implies by (2) that

$$(6) \qquad\qquad \lim_{t \to \infty} F(t) = p(\infty) \int_0^\infty g(s)\, ds$$

so long as the integral converges. This will certainly happen if

$$\int_0^\infty \mu(s, \infty]\, ds < \infty,$$

which is equivalent to the condition that $p(\infty) > 0$. It follows that, if $p(\infty) > 0$, then

$$(7) \qquad\qquad \pi(E) = \lim_{t \to \infty} P_t(x, E)$$

$$= p(\infty)\left\{ \epsilon_x(E) + \int_0^\infty g(t, E)\, dt \right\}$$

exists, and is a probability measure on S.

The conclusion is false when $p(\infty) = 0$, as can be seen by constructing functions satisfying (5) which oscillate at infinity.

(iv) Another way of expressing the conclusion of Theorem 7.4 is to say that, for fixed x, the measures $P_t(x, \cdot)$ for different values of t are mutually absolutely continuous.

(v) *A pathological example.* To see what can go wrong if the state space is not required to have a reasonably simple measure-theoretic structure, consider the following example. The state space S is the interval $[0, \infty)$, Λ is the σ-algebra of Lebesgue measurable sets in S, and m is Lebesgue measure. It is known [55] that there exists a translation invariant extension $\mathfrak{m}$ of m to a strictly larger σ-algebra Σ. Define a transition function on (S, Σ) by

$$P_t(0, E) = e^{-t}\epsilon_0(E) + \int_{E \cap (0,t)} e^{-(t-u)}\mathfrak{m}(du),$$

$$P_t(x, E) = \epsilon_{x+t}(E) \qquad (x > 0);$$

it is easy to verify that this satisfies the conditions (i), (ii) and (iii) of §7.1. Suppose if possible that there exists a function $g(t, E)$, measurable in t and defining a measure on (S, Σ) for fixed t, such that

$$P_t(0, E) = p(t)\epsilon_0(E) + \int_0^t p(t - s)g(s, E)\, ds,$$

where

$$p(t) = P_t(0, \{0\}) = e^{-t}.$$

Then, for $E \in \Sigma$, $E \not\ni 0$,

$$\int_0^t e^{-(t-s)} g(s, E)\, ds = \int_{E \cap (0,t)} e^{-(t-u)} \mathfrak{m}(du).$$

In particular, taking E to be a rational interval I,

$$\int_0^t e^s g(s, I)\, ds = \int_{I \cap (0,t)} e^u\, du,$$

which shows that, for each I,

$$g(t, I) = \epsilon_t(I)$$

for almost all t. Since there are only countably many I, there exists $N \in \Lambda$ with $m(N) = 0$, and

$$g(t, I) = \epsilon_t(I) \qquad (t \notin N).$$

But $g(t, \cdot)$ is a measure, so that we can choose a decreasing sequence (I_n) of rational intervals whose intersection is $\{t\}$, and then

$$g(t, \{t\}) = \lim g(t, I_n) = 1 \qquad (t \notin N).$$

Similarly, we may write $S - \{t\}$ as a countable union of disjoint rational intervals J_n, and then

$$g(t, S - \{t\}) = \Sigma g(t, J_n) = 0.$$

It follows that, for all $t \notin N$, $E \in \Sigma$,

$$g(t, E) = \epsilon_t(E).$$

Taking $E \in \Sigma - \Lambda$, we obtain a contradiction with the assumption that $g(\cdot, E)$ is (Lebesgue) measurable.

(vi) *A non-polish analysis.* Despite the last example, it is possible to go some way without the polish assumption, and this is done in [45]. Theorems 7.1, 7.2, 7.3, 7.4 remain valid, except for the assertion that $g(t, \cdot)$ is a measure. The proof of Theorem 7.4 requires non-trivial refinement.

It is possible that Theorem 7.5 is still true, but so far as it is at present known, only the weaker assertion can be made, that if μ is the canonical measure of the p-function of a state in a purely discontinuous process, then μ is absolutely continuous on $(0, \infty)$, and $\mu(I) > 0$ for every open interval I.

APPENDIX

Weak Convergence of Measures

The theory of weak convergence of sequences of measures on topological spaces has in recent years become an important tool in stochastic analysis, and excellent general accounts (such as that by Billingsley [2]) exist. The uses made of it in this book have however a rather limited setting, since they all relate to measures on compact intervals of the real line (including points at infinity). It may therefore be useful briefly to review the theory for this special case, and then to describe the way in which its general theorems are used to justify the various conclusions for which, in preceding chapters, the argument of weak convergence has been invoked.

Let I be a fixed interval of the form $[a, b]$, where $-\infty \leqslant a < b \leqslant \infty$, so that I is a compact space in the usual topology. Let C be the class of continuous real-valued functions on I, and M the class of finite (positive Borel) measures on I. We give M a topology (the topology of weak convergence, though we shall consider no other) defined to be the weakest topology with the property that, for every $f \in C$, the function from M into R given by

$$(1) \qquad \mu \to \int_I f(x)\mu(dx)$$

is continuous. Thus a sequence of measures μ_n converges to a measure μ if and only if

$$(2) \qquad \lim_{n \to \infty} \int_I f(x)\mu_n(dx) = \int_I f(x)\mu(dx)$$

for every $f \in C$.

The topology is clearly Hausdorff, since

$$\int_I f(x)\mu(dx) = \int_I f(x)\mu'(dx)$$

170

for all $f \in C$ only if $\mu = \mu'$. It is possible, as in [2], to go on to establish many other properties of the topological space M, but the crucial fact is that important subsets of M are compact.

Theorem A.1. For any m $(0 < m < \infty)$, the subset M_m defined by

$$(3) \qquad\qquad M_m = \{\mu \in M; \mu(I) \leqslant m\}$$

is compact.

Proof. Let $X = [0, m]^C$ be the space of functions from C into $[0, m]$, with the product topology, which by Tychonov's theorem is compact. Define a function $\gamma: M_m \to X$ by

$$(4) \qquad\qquad (\gamma\mu)(f) = \int_I f(x)\mu(dx) \qquad (f \in C).$$

Then γ is one-to-one, and identifies M_m with a subset $\gamma(M_m)$ of X. Moreover, the topology induced on $\gamma(M_m)$ from that of M_m is the same as that which $\gamma(M_m)$ inherits as a subspace of X. To prove M_m compact, it therefore suffices to show that $\gamma(M_m)$ is closed in X.

Suppose therefore that ξ lies in the closure of $\gamma(M_m)$ in X. For any $\mu \in M_m$, $\alpha, \beta \in R$, $f, g \in C$,

$$(\gamma\mu)(\alpha f + \beta g) = \alpha(\gamma\mu)(f) + \beta(\gamma\mu)(g),$$

and hence

$$\xi(\alpha f + \beta g) = \alpha\xi(f) + \beta\xi(g).$$

Moreover, if $f \geqslant 0$,

$$(\gamma\mu)(f) \geqslant 0,$$

so that

$$\xi(f) \geqslant 0.$$

Hence ξ is a positive linear functional on C, and the Riesz representation theorem [54] shows that there exists $\mu \in M$ with

$$\xi(f) = \int_I f(x)\mu(dx)$$

for all $f \in C$. In particular,

$$\mu(I) = \xi(1) \in [0, m],$$

so that $\mu \in M_m$ and $\xi = \gamma(\mu)$. Hence $\xi \in \gamma(M_m)$, and thus $\gamma(M_m)$ is closed in X, as required.   ◆

In particular, every sequence (μ_n) of measures with $\mu_n(I)$ bounded has a convergent subsequence. A further consequence deserves the status of a theorem.

Theorem A.2. *Let $F \subseteq C$ have the property that, whenever two measures μ', $\mu'' \in M$ have*

(5) $$\int_I f(x)\mu'(dx) = \int_I f(x)\mu''(dx),$$

for all $f \in F$, then $\mu' = \mu''$. Let (μ_n) be a sequence of measures for which

(6) $$\sup_n \mu_n(I) < \infty,$$

and suppose that

(7) $$\lim_{n \to \infty} \int_I f(x)\mu_n(dx)$$

exists for all $f \in F$. Then

(8) $$\mu = \lim_{n \to \infty} \mu_n$$

exists in M, and hence

(9) $$\lim_{n \to \infty} \int_I f(x)\mu_n(dx) = \int_I f(x)\mu(dx)$$

for all $f \in C$.

Proof. Because of (6), there is a value of m with $\mu_n \in M_m$ for all n, and since M_m is compact, there is a subsequence $(\mu_n; n \in \Sigma)$ converging to a limit μ_Σ. Since the limit (7) exists, it must be equal to

$$\lim_{n \in \Sigma} \int_I f(x)\mu_n(dx) = \int_I f(x)\mu_\Sigma(dx).$$

If $(\mu_n; n \in \Sigma')$ is any other convergent subsequence, this means that

$$\int_I f(x)\mu_\Sigma(dx) = \int_I f(x)\mu_{\Sigma'}(dx),$$

and the assumption made about F shows that $\mu_\Sigma = \mu_{\Sigma'}$.

Hence there is a measure μ such that every convergent subsequence of (μ_n) converges to μ. Suppose, to obtain a contradiction, that μ_n does not converge to μ. Then there exists $g \in C$, $\delta > 0$ and an infinite sequence Σ_0 of integers such that

(10)
$$\left| \int_I g(x)\mu_n(dx) - \int_I g(x)\mu(dx) \right| \geqslant \delta$$

for $n \in \Sigma_0$. By Theorem A.1, $(\mu_n; n \in \Sigma_0)$ has a convergent subsequence $(\mu_n; n \in \Sigma_1)$, and by what has already been proved, the limit of this subsequence must be μ. This contradicts (10), and the contradiction completes the proof. ◆

We now go on to describe the application of these two theorems to the various arguments of earlier chapters which depend on them. It will be seen that in several places a certain twist is needed to bring the general results to bear.

Theorem 1.6 (*page* 11)

In the proof of the lemma for this theorem we construct, for each $r \in (0, 1)$, a probability measure γ_r on $[0, \pi]$ such that

(11)
$$u_n r^n = \int_{[0,\pi]} \cos n\theta \; \gamma_r(d\theta).$$

Applying Theorem A.2 with (μ_n) replaced by (γ_r), $I = [0, \pi]$ and

$$F = \{\cos n\theta; n = 0, 1, 2, \ldots\}$$

we see that

$$\gamma = \lim_{r \to 1} \gamma_r$$

exists, and that

$$u_n = \int_{[0,\pi]} \cos n\theta \; \gamma(d\theta).$$

The fact that F satisfies the condition of the theorem is a familiar fact about Fourier–Stieltjes transforms, and (6) holds since $\gamma_r(I) = 1$. The only complication is that we are not strictly dealing with a sequence but with a collection of measures indexed by a real parameter r, but Theorem A.2 obviously extends to cover this case.

The lemma also uses the argument that, since for any $\delta > 0$, γ_r has a density g_r with

$$|g_r(\theta)| \leqslant M_\delta,$$

γ has a density on $(0, \pi)$. To prove this, note that, if $f \in C$ is non-negative and vanishes off $[\alpha, \beta] \subseteq [\delta, \pi]$, then

$$\int_I f(\theta)\gamma_r(d\theta) = \int_{[\alpha,\beta]} f(\theta)g_r(\theta)\, d\theta$$

$$\leqslant (\beta - \alpha)(\sup f)M_\delta,$$

and thus

(12) $$\int_I f(\theta)\gamma(d\theta) \leqslant (\beta - \alpha)(\sup f)M_\delta.$$

By taking functions f which are equal to 1 on intervals of the form $[\alpha + \epsilon, \beta - \epsilon]$, it follows that

$$\gamma(\alpha, \beta) \leqslant (\beta - \alpha)M_\delta,$$

which is enough to show that γ is absolutely continuous on $(0, \pi)$.

Theorem 2.8 (*page 44*)

Here we have probability measures ϕ_h on $I = [0, \infty]$ such that, if

$$n = [t/h], \qquad u = t - nh, \qquad v = 1 - u,$$

then

(13) $$p(u)\int_{[0,\infty]} p(nh - \delta)\phi_h(ds) \leqslant p^0(t)$$

$$\leqslant p(v)^{-1}\int_{[0,\infty]} p((n + 1)h - s)\phi_h(ds),$$

where $p(\tau) = 0$ for $\tau < 0$. By Theorem A.1, there is a sequence $(h(m))$ of positive numbers tending to zero, such that

(14) $$\phi = \lim_{m \to \infty} \phi_{h(m)}$$

exists.

For any $\eta > 0$, define a function $f \in C$ by

$$\begin{aligned} f(s) &= p(t - s) & (0 \leqslant s \leqslant t) \\ &= 1 - \eta^{-1}(s - t) & (t < s \leqslant t + \eta) \\ &= 0 & (t + \eta < s). \end{aligned}$$

Then

$$\lim_{m \to \infty} \int_{[0,t+\eta]} f(s)\phi_{h(m)}(ds) = \int_{[0,t+\eta]} f(s)\phi(ds),$$

so that

$$\int_{[0,t]} p(t-s)\phi(ds) + \int_{(t,t+\eta)} [1 - \eta^{-1}(s-t)]\phi(ds)$$

$$= \lim_{m \to \infty} \int_{[0,t+\eta]} f(s)\phi_{h(m)}(ds)$$

$$\geqslant \limsup_{m \to \infty} \int_{[0,t]} p(t-s)\phi_{h(m)}(ds).$$

Since η is arbitrary, this shows that

$$\limsup_{m \to \infty} \int_{[0,t]} p(t-s)\phi_{h(m)}(ds) \leqslant \int_{[0,t]} p(t-s)\phi(ds).$$

(Notice that we cannot write this down directly from the definition of weak convergence, since the integrand is not continuous on I.)

Since p is uniformly continuous on $[0, \infty)$, and since

$$|t - (n+1)h| = v \leqslant h,$$

it follows that

$$p^0(t) \leqslant \limsup_{m \to \infty} p(v)^{-1} \int_{[0,\infty]} p((n+1)h - s)\phi_h(ds)$$

$$\leqslant \int_{[0,t]} p(t-s)\phi(ds).$$

A similar argument gives the reverse inequality

$$p^0(t) \geqslant \int_{[0,t)} p(t-s)\phi(ds),$$

as required.

Theorem 3.1 (*page 59*)

This is a straightforward application of Theorem A.2, with $I = [0, \infty]$ and F consisting of the functions $k(\cdot, \theta)$ $(\theta > 0)$.

Theorem 3.2 (*page 62*)

Another direct application, with the same I and F.

Theorem 3.6 (*page* 68)

The same, yet again.

Theorem 5.2 (*page* 109)

There are two appeals to weak convergence in this proof. The first is just the same as that used in proving Theorem 3.1, except that F is replaced by the slightly smaller class $\{k(\cdot, \theta); \theta \notin \mathscr{L}\}$. As remarked, however, the necessary property of F, that

$$\int_I f(x)\mu'(dx) = \int_I f(x)\mu''(dx)$$

for all $f \in F$ should imply $\mu' = \mu''$, still obtains.

In the second use of the argument, we have the equation

$$(15) \qquad -q_{ij}(\theta) = \lim_{h \to 0} \int e^{-\theta x} \lambda_{ij}^{(h)}(dz),$$

for all $\theta > 0$, $\theta \notin \mathscr{L}$, where $\lambda_{ij}^{(h)}$ is a totally finite measure on $[0, \infty)$, and we have to establish the existence of a measure λ_{ij} on $[0, \infty)$ such that

$$-q_{ij}(\theta) = \int e^{-\theta x} \lambda_{ij}(dx).$$

Since i and j are fixed throughout the argument, we omit them to simplify the notation.

Take any $\alpha > 0$, $\alpha \notin \mathscr{L}$ and define a finite measure λ_α^h on $I = [0, \infty]$ by

$$\lambda_\alpha^h(E) = \int_{E \cap [0, \infty)} e^{-\alpha x} \lambda^{(h)}(dx).$$

Then (15) shows that

$$-q(\alpha) = \lim_{h \to 0} \lambda_\alpha^h(I),$$

so that the measures λ_α^h have bounded total mass as $h \to 0$.

Apply Theorem A.2 to these measures, with F consisting of the functions

$$e^{-(\theta - \alpha)x} \qquad (\theta \geqslant \alpha, \theta \notin \mathscr{L}).$$

We have first to check that F satisfies the condition of the theorem. To do this, let μ' and μ'' be finite measures on $I = [0, \infty]$ with

$$\int e^{-(\theta - \alpha)x} \mu'(dx) = \int e^{-(\theta - \alpha)x} \mu''(dx)$$

for all $\theta \geqslant \alpha$, $\theta \notin \mathscr{L}$. By continuity this holds for all $\theta \geqslant \alpha$, and by the uniqueness theorem for Laplace–Stieltjes transforms $\mu' = \mu''$.

Theorem A.2 therefore shows that

$$\lambda_\alpha = \lim_{h \to 0} \lambda_\alpha^h$$

exists as a finite measure on $[0, \infty]$, and then (15) implies that

(16)
$$-q(\theta) = \int_{[0,\infty]} e^{-(\theta-\alpha)x} \lambda_\alpha(dx)$$

whenever $\theta \geqslant \alpha$, $\theta \notin \mathscr{Z}$. It follows that $-q(\theta)$ is non-increasing and continuous in $\theta > \alpha$, and since α can be taken arbitrarily small (since $\mathscr{Z}$ is countable), $-q(\theta)$ is non-increasing and continuous in $\theta > 0$. Since

$$\lambda_\alpha(\{\infty\}) = -q(\alpha) + \lim_{\theta \to \alpha} q(\theta),$$

it follows that $\lambda_\alpha(\{\infty\}) = 0$. We may therefore define a measure $\lambda_{(\alpha)}$ on $[0, \infty)$ by

$$\lambda_{(\alpha)}(E) = \int_E e^{\alpha x} \lambda_\alpha(dx).$$

Equation (16) shows that

$$-q(\theta) = \int_{[0,\infty)} e^{-\theta x} \lambda_{(\alpha)}(dx)$$

for $\theta \geqslant \alpha$, $\theta \notin \mathscr{Z}$. Hence $\lambda_{(\alpha)}$ does not depend on α, but is a measure λ with

$$-q(\theta) = \int_{[0,\infty)} e^{-\theta x} \lambda(dx)$$

for $\theta > 0$, $\theta \notin \mathscr{Z}$.

Notice that there is no way of arguing that λ is totally finite; this comes later in the proof.

This particular argument can be evaded by appeal to deep facts about Laplace–Stieltjes transforms [73].

Theorem 7.2 (*page* 157)

This differs only notationally from the argument already described in relation to Theorem 2.8.

Bibliography

1. M. S. Bartlett, "Recurrence and first passage times", *Proc. Cam. Phil. Soc.* **49** (1953), 263–275.
2. P. Billingsley, *Convergence of Probability Measures*, Wiley, New York (1968).
3. N. H. Bingham, "Limit theorems for regenerative phenomena, recurrent events and renewal theory", *Ztschr. Wahrsch'theorie & verw. Geb.*
4. D. Blackwell and D. Freedman, "On the local behaviour of Markov transition probabilities", *Ann. Math. Statist.* **39** (1968), 2123–2127.
5. P. Bloomfield, "Lower bounds for renewal sequences and p-functions", *Ztschr. Wahrsch'theorie & verw. Geb.* **19** (1971), 271–273.
6. K. L. Chung, *Markov Chains with Stationary Transition Probabilities*, Springer, Berlin (1967).
7. D. R. Cox, *Renewal Theory*, Methuen, London (1962).
8. H. T. Croft, "A question of limits", *Eureka* **20** (1957), 11–13.
9. D. J. Daley, "On a class of renewal functions", *Proc. Cam. Phil. Soc.* **61** (1965), 519–526.
10. ——, Contribution to the discussion of 42.
11. R. Davidson, Ph.D. thesis (unpublished), University of Cambridge (1968).
12. ——, "Arithmetic and other properties of certain delphic semigroups", *Ztschr. Wahrsch'theorie & verw. Geb.* **10** (1968), 120–172.
13. ——, "More delphic theory and practice", *Ztschr. Wahrsch'theorie & verw. Geb.* **13** (1969), 191–203.
14. W. Doeblin, "Sur deux problèmes de M. Kolmogoroff concernant les chaines dénomerables", *Bull. Soc. math. Fr.* **66** (1938), 210–220.
15. J. L. Doob, *Stochastic Processes*, Wiley, New York (1953).
16. P. Erdös, W. Feller and H. Pollard, "A theorem on power series", *Bull. Am. math. Soc.* **55** (1949), 201–204.
17. W. Feller, "Fluctuation theory of recurrent events", *Trans. Am. math. Soc.* **67** (1949), 98–119.
18. ——, *An Introduction to Probability Theory and its Applications*, Wiley, New York (1950).
19. ——, *An Introduction to Probability Theory and its Applications, Volume II*, Wiley, New York (1966).

179

20. W. Feller and S. Orey, "A renewal theorem", *J. Math. Mech.* **10** (1961), 619–624.
21. D. Freedman, *Markov Chains*, Holden-Day, San Francisco (1971).
22. P. R. Halmos, *Measure Theory*, van Nostrand, Princeton (1950).
23. G. H. Hardy, J. E. Littlewood and G. Polya, *Inequalities*, Cambridge University Press, Cambridge (1934).
24. E. Hille and R. S. Phillips, *Functional Analysis and Semi-groups*, American Mathematical Society, Providence (1957).
25. J. Hoffmann–Jørgensen, "Markov sets", *Mathematica scand.* **24** (1969), 145–66.
26. T. Kaluza, "Über die Koeffizienten reziproker Potenzreihen", *Math. Z.* **28** (1928), 161–170.
27. J. L. Kelley, *General Topology*, van Nostrand, Princeton (1955).
28. D. G. Kendall, "Some analytical properties of continuous stationary Markov transition functions", *Trans. Am. math. Soc.* **78** (1955), 529–540.
29. ——, "Some problems in the theory of dams", *J. R. statist. Soc. B.* **19** (1957), 207–233.
30. ——, "On the behaviour of a standard Markov transition function near $t = 0$", *Ztschr. Wahrsch'theorie & verw. Geb.* **3** (1965), 276–278.
31. ——, "Some recent developments in the theory of denumerable Markov processes", *Transactions of the 4th Prague conference on information theory, statistical decision functions and random processes*, Prague (1967).
32. ——, "Renewal sequences and their arithmetic", *Proceedings of the Loutraki symposium on probability*, Springer, Berlin (1967).
33. ——, "Delphic semi-groups, infinitely divisible regenerative phenomena, and the arithmetic of p-functions", *Ztschr. Wahrsch'theorie & verw. Geb.* **9** (1968), 163–195.
34. J. F. C. Kingman, "The imbedding problem for finite Markov chains", *Ztschr. Wahrsch'theorie & verw. Geb.* **1** (1962), 14–24.
35. ——, "The exponential decay of Markov transition probabilities", *Proc. Lond. math. Soc.* **13** (1963), 337–358.
36. ——, "Ergodic properties of continuous-time Markov processes and their discrete skeletons", *Proc. Lond. math. Soc.* **13** (1963), 593–604.
37. ——, "A continuous time analogue of the theory of recurrent events", *Bull. Am. math. Soc.* **69** (1963), 268–272.
38. ——, "The stochastic theory of regenerative events", *Ztschr. Wahrsch'-theorie & verw. Geb.* **2** (1964), 180–224.
39. ——, "Linked systems of regenerative events", *Proc. Lond. math. Soc.* **15** (1965), 125–150.
40. ——, "On doubly stochastic Poisson processes", *Proc. Cam. Phil. Soc.* **60** (1964), 923–930.
41. ——, "Some further analytical results in the theory of regenerative events", *J. math. Analysis Applic.* **11** (1965), 422–433.
42. ——, "An approach to the study of Markov processes", *J. R. statist. Soc. B.* **28** (1966), 417–447.
43. ——, "Markov transition probabilities I", *Ztschr. Wahrsch'theorie & verw. Geb.* **6** (1967), 248–270.
44. ——, "Markov transition probabilities II; completely monotonic functions", *Ztschr. Wahrsch'theorie & verw. Geb.* **9** (1967), 1–9.

45. ——, "Markov transition probabilities III; general state spaces", *Ztschr. Wahrsch'theorie & verw. Geb.* **10** (1968), 87–101.
46. ——, "Markov transition probabilities IV; recurrence time distributions", *Ztschr. Wahrsch'theorie & verw. Geb.* **11** (1968), 9–17.
47. ——, "Markov transition probabilities V", *Ztschr. Wahrsch'theorie & verw. Geb.* **17** (1971), 89–103.
48. ——, "On measurable p-functions", *Ztschr. Wahrsch'theorie & verw. Geb.* **11** (1968), 1–8.
49. ——, "A class of positive-definite functions", *Problems in Analysis (ed. R. C. Gunning)*, Princeton University, Princeton (1970).
50. ——, "Stationary regenerative phenomena", *Ztschr. Wahrsch'theorie & verw. Geb.* **15** (1970), 1–18.
51. ——, "An application of the theory of regenerative phenomena", *Proc. Cam. Phil. Soc.* **68** (1970), 697–701.
52. ——, "Regenerative phenomena and the characterisation of Markov transition probabilities", *Proceedings of the 6th Berkeley symposium on mathematical statistics and probability*, University of California Press, Berkeley (to appear).
53. J. F. C. Kingman and S. Orey, "Ratio limit theorems for Markov chains", *Proc. Am. math. Soc.* **15** (1964), 907–910.
54. J. F. C. Kingman and S. J. Taylor, *Introduction to Measure and Probability*, Cambridge University Press, Cambridge (1966).
55. K. Kodaira and S. Kakutani, "A non-separable translation invariant extension of the Lebesgue measure space", *Ann. Math.* **52** (1950), 574–579.
56. A. N. Kolmogorov, "Anfangsgünde der Theorie der Markoffschen Ketten mit unendlichen vielen möglichen Zuständen", *Mat. Sbornik* (1936) 607–610.
57. P. Lévy, "Systèmes markoviens et stationnaires; cas dénomerable", *Ann. scient. Éc. norm. sup. Paris* **68** (1951), 327–381.
58. M. Loève, *Probability Theory*, van Nostrand, Princeton (1963).
59. R. M. Loynes, "On certain applications of the spectral representation of stationary processes", *Ztschr. Wahrsch'theorie & verw. Geb.* **5** (1966), 180–186.
60. E. J. McShane, *Integration*, van Nostrand, Princeton (1944).
61. J. Neveu, "Une généralisation des processus à accroissements positifs indépendants", *Abhandlungen aus dem Mathematischen Seminar, Universität Hamburg* **25** (1961), 36–61.
62. S. Orey, "Strong ratio limit property", *Bull. Am. math. Soc.* **67** (1961), 571–574.
63. D. Ornstein, "The differentiability of transition functions", *Bull. Am math. Soc.* **66** (1960), 36–39.
64. G. E. H. Reuter, "Über eine Volterrasche Integralzleichung mit total-monotonem Kern", *Arch. Math.* **7** (1956), 59–66.
65. M. Rubinovitch, "Ladder phenomena in stochastic processes with stationary independent increments", *Ztschr. Wahrsch'theorie & verw. Geb.* **20** (1971), 58–74.
66. S. M. Rudolfer, "Some metric invariants for Markov shifts", *Ztschr. Wahrsch'theorie & verw. Geb.* **15** (1970), 202–207.
67. G. J. Smith, "Instantaneous states of Markov processes", *Trans Am. math. Soc.* **110** (1964), 185–195.

68. W. L. Smith, "Regenerative stochastic processes", *Proc. R. Soc. A.* **232** (1955), 6–31.
69. ——, "Renewal theory and its ramifications", *J. R. statist. Soc. B.* **20** (1958), 243–302.
70. A. M. Sykes, D. Phil. thesis (unpublished), University of Sussex (1970).
71. L. Takacs, *Combinational Methods in the Theory of Stochastic Processes*, Wiley, New York (1967).
72. D. Vere-Jones, "Geometric ergodicity in denumerable Markov chains", *Q. J. Math.* **13** (1962), 7–28.
73. D. V. Widder, *The Laplace Transform*, Princeton University, Princeton (1941).
74. D. V. Widder and I. J. Hirschman, *The Convolution Transform*, Princeton University, Princeton (1955).
75. T. Yamada, "$\mathscr{E}$-regenerative phenomena in some stochastic processes", *Kodai mathematical seminar reports* **20** (1968), 76–93.
76. A. A. Yuskevitch, "On differentiability of transition probabilities of homogeneous Markov processes with a countable number of states", [Russian] *Uchenye zapiski MGU* **186** (1959), 141–159.

Supplementary Bibliography

While this book was in the hands of the printer, further progress had been made with the problems of regenerative phenomena, and some new papers are listed here with other relevant material. It is fitting that several of these are to be published (by John Wiley & Sons Ltd.), in the volume abbreviated as *SGSA*: *Stochastic Geometry and Stochastic Analysis*, edited by E. F. Harding and D. G. Kendall in memory of Rollo Davidson. In particular, this includes a posthumous paper of Davidson [80] in which the inequality (3.5.14) is sharpened to $\vartheta_0 \leqslant \frac{2}{3}$ (a striking vindication of the last sentence of the Preface). Kendall's contributions fulfil the hope expressed on page 82, and, as a by-product of a more general study, provide a framework in which the development of Chapter 5 may systematically be fitted.

Increasing interest is now being shown in what I have here called fictitious regenerative phenomena, and in [93] and [83] this problem is attacked from a basis laid by a paper of Krylov and Yuskevitch [92]. Such phenomena present rather different challenges from those of the ordinary theory, and the remarks at the end of Chapter 4 are still pertinent. A measure of the difficulties can be gained from the important paper of Kesten [90].

Another generalisation is suggested by recent work of Cornish [79], who follows Jurkat [84] [85] in studying arrays of functions p_{ij} satisfying (2.1.3) but not necessarily the row-sum condition in (2.1.2). There is an appropriate generalisation of the theory of p-functions, described in [91].

77. N. H. Bingham, "Limit theorems for a class of Markov processes", *SGSA*.
78. P. Bloomfield, "Stochastic inequalities for regenerative phenomena", *SGSA*.
79. A. G. Cornish, "Some properties of semi-groups of non-negative matrices and their resolvents", *Proc. Lond. math. Soc.* (to appear)
80. R. Davidson, "Smith's phenomenon and 'jump' p-functions", *SGSA*.
81. —— and D. G. Kendall, "On partly exponential p-functions and 'identifying' skeletons", *SGSA*.
82. J. Horowitz, "A note on the arc-sine law and Markov random sets", *Ann. math. Statist.* **42** (1971), 1068–1074.
83. ——, "Semilinear Markov processes, subordinators and renewal theory", *Ztschr. Wahrsch'theorie & verw. Geb.*

84. W. B. Jurkat, "On semi-groups of positive matrices", *Scripta Math.* **24** (1959), 123–131 and 207–218.

85. ——, "On the analytic structure of semi-groups of positive matrices", *Math. Z.* **73** (1960), 346–365.

86. D. G. Kendall, "Stochastic analysis and stochastic geometry", *SGSA*.

87. ——, "Separability and measurability for stochastic processes: a survey", *SGSA*.

88. ——, "Foundations of a theory of random sets", *SGSA*.

89. ——, "On the non-occurrence of a regenerative phenomenon in given intervals", *SGSA*.

90. H. Kesten, "Hitting probabilities of single points for processes with stationary independent increments", *Mem. Amer. Math. Soc.* **93** (1969).

91. J. F. C. Kingman, "Semi-p-functions".

92. N. V. Krylov and A. A. Yuskevitch, "Markov random sets", *Teor. Verojatnost.* **9** (1964), 738–743.

93. P. A. Meyer, "Ensembles régéneratifs, d'après Hoffmann–Jørgensen", *Seminaire de Probabilitès* (Strasbourg), IV, 133–150.

Index of Theorems

Author Index

187

Subject Index